ELEMENTARY ANALYSIS

VOLUME 1

ELEMENTARY ANALYSIS

VOLUME 1

BY

K. S. SNELL AND J. B. MORGAN

PERGAMON PRESS

OXFORD · LONDON · EDINBURGH · NEW YORK

PARIS · FRANKFURT

Pergamon Press Ltd., Headington Hill Hall, Oxford
4 & 5 Fitzroy Square, London W.1

Pergamon Press (Scotland) Ltd., 2 & 3 Teviot Place, Edinburgh 1

Pergamon Press of Canada Ltd., 6 Adelaide Street East, Toronto, Ontario

Pergamon Press Inc., 44–01 21st Street, Long Island City, New York 11101

Pergamon Press S.A.R.L., 24 rue des Écoles, Paris 5e

Pergamon Press GmbH, Kaiserstrasse 75, Frankfurt-am-Main

First edition 1965
Library of Congress Catalog Card No. 65-25335

Set in 10 on 12 pt. Times and Printed in Great Britain by Page Bros. (Norwich) Ltd.

(1797/65)

CONTENTS

PREFACE

THE two volumes of *Elementary Analysis* are intended to introduce many of the ideas of modern mathematics in an informal manner, and also to provide the practical experience in algebraic and analytic operations which will lay a sound foundation of basic skills. They are intended for use in the top forms of schools and in college courses, both academic and technical.

Amongst the important ideas developed in the two volumes are the nature of number, algebraic and logical structure, groups, rings, fields, vector spaces, matrices, sequences, limits, functions and inverse functions, complex numbers and probability. A glimpse into the logical structure of analysis is given through the treatment of differentiation and integration, with applications to the trigonometric and logarithmic functions.

In the main the preliminary ideas are dealt with in Volume 1 and illustrated by applications to the simpler algebraic functions, although there are important exceptions such as the theory of the exponential function and its inverse function. Volume 2 begins with a description of the trigonometric functions of the general angle which can be taken as soon after Chapter 3 as desired; it also contains an introduction to the binomial theorem and series which can be read immediately after Chapter 10, or even after Chapter 5.

The latter half of Volume 2, from Chapter 23 (Numerical Solution of Equations) onwards, need not be read in the order in which the chapters occur, although a student with only a little knowledge of analytical geometry (Chapter 24) might need some further acquaintance with it before reading Chapter 25 (The Argand Diagram). Chapter 23 is itself rather a feature of Volume 2 since it deals in some detail with numerical methods and methods of approximation that form such an important section of modern applied mathematics.

The underlying philosophy of *Elementary Mathematics* is that mathematics is basically a double-sided study, both scientific and artistic: it is a science of logical thought, a study of axiomatics and structure; it is also the art of solving problems, of creating relations between practical and numerical systems and using the properties of the numerical systems to discover meaningful and helpful solutions. The authors believe that the educational value of mathematics lies in the interplay between these two characteristics, and their belief has governed the growth of this book.

The book aims at giving an understanding of fundamental principles, rather than manipulative techniques. For this reason it is of particular value to teachers and students in training colleges; it should help teachers to see how well-chosen graphical work in an elementary course can make a good foundation for further development.

There are examples in the text for the student to do as he reads the text, as well as full sets of exercises. The former are meant to help understanding of each section. Answers are given to all but the easiest questions.

K. S. S.

May 1965 J. B. M.

CHAPTER 1

NUMBER SYSTEMS

Number scales

The early steps in the evolution of numbers are lost in the silence of pre-history, but studies of primitive tribes have suggested ways in which number systems may have been developed. The needs of barter, calendar making, military operations, and the administration of early civilizations, gave rise to methods of counting from which written number systems were eventually abstracted.

The simplest numbers are those used for counting. Counting on fingers and toes led to the use of basic groups of 5, 10 or 20, but 12 and 60 were also used. The most common *number base* is ten, and we are accustomed to writing our numbers in the *denary scale*, a method invented by the Hindus possibly as early as the third century B.C.; the zero symbol was a later invention, but certainly not later than the ninth century. The complete number system was explained by the Arab mathematician *al-Khwarizmi* in a work written about 820. The first European to write a complete account was Fibonacci, in his *Liber Abaci*, which appeared in 1202, but it was nearly a century later before the new system was generally accepted.

Each number has both a name, such as zero, one, two three, . . . and a symbol, 0, 1, 2, 3, . . . The advantage of the Hindu–Arabic system lies in the use of the ten digits to represent different numbers according to the positions in which they are written; for example, in 345 the 3 represents 3 hundreds, the 4 represents 4 tens and the 5 represents 5 units, so that

$$345 = 3.10^2 + 4.10 + 5.$$

Similarly, $10326 = 1.10^4 + 0.10^3 + 3.10^2 + 2.10 + 6$. The importance of 0, the zero symbol, and the purpose served by 10, the base, are clearly shown by this example.

However, we can use any number as the base. Since $142 = 1.3^4 + 2.3^3 + 0.3^2 + 2.3 + 1$, the denary scale number 142 can be written as 12021 in the scale of three, or the ternary scale.

The system in which two is the base is called the *binary scale*; only the two digits 0 and 1 are required, and they are often called *bits*, a contraction of *binary digits*. The binary scale is of particular importance because of the simplicity with which it can be represented electrically. For example, if a lamp is used to represent a binary digit, it can represent 0 when switched off and 1 when switched on. This is the reason for the use of the binary scale in electronic digital computers.

In the binary scale,

$$1011011 = 1.2^6 + 0.2^5 + 1.2^4 + 1.2^3 + 0.2^2 + 1.2 + 1.$$

Notice that the binary digits are given by the successive remainders on repeated division by 2, and this will be true in whatever scale the original number is given. This last number is 91 in the denary scale, and repeated division of 91 by 2 gives:

2	91	
2	45	. . . 1
2	22	. . . 1
2	11	. . . 0
2	5	. . . 1
2	2	. . . 1
	1	. . . 0

The arrows show the order in which the binary digits are given, 1011011.

Ex. 1. Express in the binary scale: 9, 12, 50.

Ex. 2. Express the following binary numbers in the denary scale: 1010, 10101, 1110111.

Ex. 3. Working in the binary scale, evaluate:

(i) $110 + 1010$, (ii) 111×11, (iii) $1010/10$.

Ex. 4. Express the following ternary numbers in the binary and denary scales: 121, 1202.

Rational numbers

In the very early civilizations of Babylon and Egypt the subdivision of units of measure, in order to provide greater accuracy in coinage, surveying and astronomy, gave rise to the systematic use of fractions, which we now call *rational numbers*. Methods of calculating with fractions were clumsy, and again the decimal system of the Hindus provided the first convenient and efficient method; thus

$$20{\cdot}49 = 2 \times 10 + 0 \times 1 + 4 \times \frac{1}{10} + 9 \times \frac{1}{10^2}$$

Similarly, by extending the method to the binary scale,

$$101{\cdot}101 = 1 \times 2^2 + 0 \times 2 + 1 \times 1 + 1 \times \frac{1}{2} + 0 \times \frac{1}{2^2} + 1 \times \frac{1}{2^3}$$

The binary digits for the fractional part of the number are exposed by repeated multiplication by 2, and again it does not matter in which scale the original number is given.

For example, working from the denary scale:

$$\tfrac{5}{8} \times 2 = 1 + \tfrac{1}{4}, \quad \tfrac{1}{4} \times 2 = 0 + \tfrac{1}{2}, \quad \tfrac{1}{2} \times 2 = 1 + 0;$$

hence, in the binary scale, $\frac{5}{8} = 0{\cdot}101$ exactly.

Alternatively, working from decimals, successive multiplications by 2 give:

$$\tfrac{5}{8} = 0{\cdot}625, \quad 2 \times 0{\cdot}625 = 1{\cdot}25, \quad 0{\cdot}25 \times 2 = 0{\cdot}5, \quad 0{\cdot}5 \times 2 = 1{\cdot}0,$$

giving $\frac{5}{8} = 0{\cdot}101$ in the binary scale.

Again

$$\tfrac{5}{7} \times 2 = 1 + \tfrac{3}{7}, \quad \tfrac{3}{7} \times 2 = 0 + \tfrac{6}{7}, \quad \tfrac{6}{7} \times 2 = 1 + \tfrac{5}{7},$$

and so on. In the binary scale,

$$\tfrac{5}{7} = 0{\cdot}101101\ldots = 0{\cdot}\dot{1}0\dot{1} \quad \text{(recurring)}.$$

Binary numbers like $0{\cdot}101$ and $0{\cdot}\dot{1}0\dot{1}$ are called *bicimal fractions*, or *bicimals*, corresponding to decimal fractions, or decimals, in the denary scale.

Ex. 5. Express as bicimals: 0·5, 0·125, 0·375.
Ex. 6. Express as recurring bicimals: 0·3, 0·7, 0·9.
Ex. 7. Express as decimal fractions the bicimal fractions: 0·11, 10·01, 1·101.
Ex. 8. Working in the binary scale, simplify:
(i) $0{\cdot}11 + 1{\cdot}011$, (ii) $1{\cdot}001 - 0{\cdot}111$, (iii) $10{\cdot}1 \times 0{\cdot}111$.

Geometrical representation of rational numbers

Figure 1.1 shows a straight line on which is chosen a fixed origin, O. Equal lengths are measured to the right of O along the line, and are labelled 1, 2, 3, ... Corresponding to each positive integer n there is a point P on the line such that $OP = n$ units.

Similarly the negative integers can be represented by points marked on the line to the left of O.

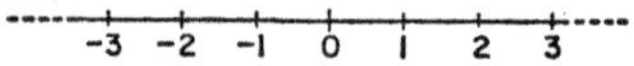

Fig. 1.1

Any vulgar fraction can be expressed in the form p/q, where p and q are integers. By subdividing each unit length on the line into q equal parts, we can represent p/q on the line by the point Q, where $OQ = p(1/q)$. All rational numbers, positive, negative or zero, can therefore be represented by points marked on the line and referred to the origin O, which itself represents zero.

If we try to represent every rational number by a decimal fraction we meet trouble at once, for $\tfrac{1}{3} = 0{\cdot}333\ldots$, a recurring decimal. This naturally raises the question whether every recurring decimal represents a rational number, and we find that it does; the proof for a particular case is applicable to any recurring decimal.

$$\begin{aligned}
\text{Suppose} \qquad x &= 3{\cdot}1\dot{2}0\dot{2},\\
10x &= 31{\cdot}\dot{2}0\dot{2},\\
10x - 31 &= 0{\cdot}\dot{2}0\dot{2},\\
10000x - 31000 &= 202{\cdot}\dot{2}0\dot{2},\\
10000x - 31202 &= 0{\cdot}\dot{2}0\dot{2} = 10x - 31,\\
9990x &= 31171,\\
\text{and so} \qquad x &= 31171/9990.
\end{aligned}$$

Ex. 9. Find the rational numbers represented by the following recurring decimals:

$$\text{(i) } 2{\cdot}\dot{2}, \quad \text{(ii) } 1{\cdot}\dot{1}\dot{2}, \quad \text{(iii) } 0{\cdot}\dot{5}0\dot{4}.$$

Ex. 10. Express the following recurring bicimals as rational numbers in the binary scale:

$$\text{(i) } 0{\cdot}\dot{1}, \quad \text{(ii) } 1{\cdot}\dot{0}\dot{1}, \quad \text{(iii) } 10{\cdot}\dot{1}0\dot{1}.$$

Nested intervals

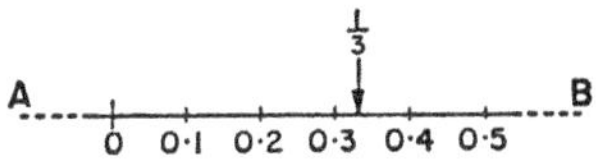

Fig. 1.2

Suppose we form a sequence of intervals on the line AB in Fig. 1.2 in the following manner:

(i) $0 \leqslant x \leqslant 1$, (ii) $0{\cdot}3 \leqslant x \leqslant 0{\cdot}4$,

(iii) $0{\cdot}33 \leqslant x \leqslant 0{\cdot}34$, (iv) $0{\cdot}333 \leqslant x \leqslant 0{\cdot}334$,

and so on.

Each interval is chosen in such a way that it lies entirely within the preceding interval. Such a sequence is called a *nest of intervals*.

Since the length of each interval is one-tenth of the length of the preceding interval, no number other than $\frac{1}{3}$ can lie in every one of the intervals if the process is continued indefinitely.

Furthermore, every rational number can be represented by a nest of intervals of this kind. If a rational number is expressed as a decimal, it produces either a *recurring decimal* or a terminating decimal. We have seen above how to deal with a recurring decimal, but a terminating decimal can be dealt with in a similar way.

For example, to represent 0·3 we construct the nested intervals:
(i) $0 \leqslant x \leqslant 1$, (ii) $0{\cdot}2 \leqslant x \leqslant 0{\cdot}3$, (iii) $0{\cdot}29 \leqslant x \leqslant 0{\cdot}30$,
(iv) $0{\cdot}299 \leqslant x \leqslant 0{\cdot}300$, (v) $0{\cdot}2999 \leqslant x \leqslant 0{\cdot}3000$,
and so on.

This suggests that $0{\cdot}2\dot{9}$ is equivalent to 0·3.

Ex. 11. Express $0{\cdot}2\dot{9}$ as a rational number.

Ex. 12. Describe a nest of intervals representing 1·25.

Irrational numbers

Every rational number can be represented by a nest of intervals which defines a specific point on the line AB of Fig. 1.2. Is the converse true? Is each point of the line defined by a nest of intervals which represents a rational number? Surprisingly the answer is no, and even more surprisingly the Pythagoreans had reached this conclusion before 500 B.C. A modern equivalent of their reasoning is given below. The symbol $\Rightarrow$ is used to indicate that *if* one statement is true, *then* the subsequent statement is true. For example,

$$x = 1 \Rightarrow x^2 = 1, \quad \text{and } x = 2 \Rightarrow x^2 > 3.$$

A double arrow $\Leftrightarrow$ is used to combine the two statements $P \Rightarrow Q$ *and* $Q \Rightarrow P$; thus

$$x > 0 \Leftrightarrow 2x > 0, \text{ and } x^2 = 1 \Leftrightarrow x = \pm 1.$$

Consider the case of $\sqrt{2}$, and suppose $\sqrt{2} = p/q$ where p and q are integers which have no common factor. Then

$$p/q = \sqrt{2} \Rightarrow p^2 = 2q^2.$$

Hence p^2 is an even number. But the square of an odd number is an odd number, and so p is even. Let $p = 2r$; then

$$4r^2 = 2q^2 \Leftrightarrow q^2 = 2r^2.$$

A repetition of the argument shows that q is also an even number, so that p and q must have a common factor 2. The supposition that $\sqrt{2} = p/q$ therefore leads to a contradiction, and so must be false. It follows that $\sqrt{2}$ is not a rational number; it is *irrational*.

Suppose we now express $\sqrt{2}$ as a decimal, for example by evaluating the numbers $x_2, x_3, \ldots$ where $0 < x_1$, $x_2 = \frac{1}{2}(x_1 + 2/x_1)$, $x_3 = \frac{1}{2}(x_2 + 2/x_2)$, and so on, which is a method suggested by Newton. The result is a succession of numbers approximating more and more closely to $\sqrt{2}$, as will be deduced from the Example below. The decimal we find will neither terminate nor recur, but we can continue the process until we have found the value of $\sqrt{2}$ to as many decimal places as we wish to find. In this way we can derive a nest of intervals to represent $\sqrt{2}$.

EXAMPLE. *Suppose* $y = \frac{1}{2}(x + 2/x)$. *Show that (i) if* $x > \sqrt{2}$, *then* $x > y > \sqrt{2}$, *and (ii) if* $0 < x < \sqrt{2}$, *then* $y > \sqrt{2}$.

(i) $x - y = \dfrac{1}{2}\left(x - \dfrac{2}{x}\right) = \dfrac{1}{2x}(x^2 - 2)$, so that $x > y$.

Also $y - \sqrt{2} = \dfrac{1}{2}\left(x - 2\sqrt{2} + \dfrac{2}{x}\right) = \dfrac{1}{2x}(x - \sqrt{2})^2$,

so that $y > \sqrt{2}$.

(ii) We have $y - \sqrt{2} = \dfrac{1}{2x}(x - \sqrt{2})^2$, and so $y > \sqrt{2}$.

Note. From (i),

$$\frac{y - \sqrt{2}}{x - \sqrt{2}} = \frac{x - \sqrt{2}}{2x} = \frac{1}{2} - \frac{1}{x\sqrt{2}}.$$

Hence $\qquad x > \sqrt{2} \Rightarrow y - \sqrt{2} < \frac{1}{2}(x - \sqrt{2}).$

If we start with $x_1 > \sqrt{2}$ and calculate the numbers $x_2, x_3, x_4, \ldots$ from Newton's formula,

$$x_2 = \tfrac{1}{2}(x_1 + 2/x_1), \quad x_3 = \tfrac{1}{2}(x_2 + 2/x_2),$$

and so on, it follows that:

$$x_2 - \sqrt{2} < \frac{1}{2}(x_1 - \sqrt{2}), \quad x_3 - \sqrt{2} < \frac{1}{2^2}(x_1 - \sqrt{2}),$$

$$x_4 - \sqrt{2} < \frac{1}{2^3}(x_1 - \sqrt{2}), \quad \text{and so on.}$$

The difference between x and $\sqrt{2}$ is therefore reduced by more than one-half of its value by each application of the formula; by applying the formula repeatedly we can find $\sqrt{2}$ to whatever degree of accuracy we wish. This method is very suitable for machine calculations.

Ex. 13. Evaluate x_3 when $x_1 = 1$, $x_2 = \frac{1}{2}(x_1 + 2/x_1)$, $x_3 = \frac{1}{2}(x_2 + 2/x_2)$.

Ex. 14. Explain briefly how to find $\sqrt{3}$ by Newton's method. How would you find $\sqrt{N}$?

Since the rational numbers which define the boundaries of each nested interval in an approximation to $\sqrt{2}$ follow the laws of arithmetic, so will the number $\sqrt{2}$ itself, to any degree of accuracy that we care to specify. Furthermore, we have now used up all the points of the line on which we represent our system of numbers, since each point of the line can be defined by a system of nested intervals.

The real numbers

The system of numbers which represents every point of the line AB in Fig. 1.2 is called the *real number* system. We have shown, with considerable appeals to intuition, that the real numbers must follow the same laws of addition, subtraction, multiplication and division as those which are followed by the rational numbers defining the nested intervals.

The rules of operation for real numbers will be briefly examined, but we shall not attempt a rigorous examination, either of the numbers or of the rules. The reader who wishes to study a strict account of such work should refer to a more advanced textbook on mathematical analysis, such as *A Course of Pure Mathematics*, G. H. Hardy.

EXAMPLE. *Express* $(2 - \sqrt{3})/(\sqrt{3} + 1)$ *in the form* $a + b\sqrt{3}$.

$$\frac{2 - \sqrt{3}}{\sqrt{3} + 1} = \frac{(2 - \sqrt{3})(\sqrt{3} - 1)}{(\sqrt{3} + 1)(\sqrt{3} - 1)}$$

$$= \frac{3\sqrt{3} - 5}{3 - 1}$$

$$= \frac{3}{2}\sqrt{3} - \frac{5}{2}.$$

Note. Numbers like $\sqrt{2}$, $3 + \sqrt{2}$, $2\sqrt{3} - 3\sqrt{2}$ are sometimes called *surds*. Numerical work can often be carried out more conveniently with surds than with their decimal approximations. For example,

$$\frac{1}{\sqrt{2}} = \frac{1}{\sqrt{2}} \times \frac{\sqrt{2}}{\sqrt{2}} = \frac{1}{2}\sqrt{2},$$

$$\frac{1}{\sqrt{2} - 1} = \frac{1}{\sqrt{2} - 1} \times \frac{\sqrt{2} + 1}{\sqrt{2} + 1} = \frac{\sqrt{2} + 1}{2 - 1} = \sqrt{2} + 1.$$

EXAMPLE. *Solve* $\sqrt{(2x - 3)} - \sqrt{(x + 10)} = 1$.

Note. When both sides of an equation are squared a possible ambiguity is introduced:

$$x = a \text{ and } x = -a \text{ both lead to } x^2 = a^2.$$

It is therefore *essential* to check solutions.

$$\sqrt{(2x - 3)} - \sqrt{(x + 10)} = 1,$$
$$\sqrt{(2x - 3)} = 1 + \sqrt{(x + 10)}.$$

Squaring both sides:

$$2x - 3 = 1 + 2\sqrt{(x + 10)} + (x + 10),$$
$$x - 14 = 2\sqrt{(x + 10)}$$

Squaring again:

$$x^2 - 28x + 196 = 4x + 40,$$
$$x^2 - 32x + 156 = 0,$$
$$(x - 6)(x - 26) = 0.$$

Hence $x = 6$ or 26.

When $x = 6$, $\sqrt{(2x - 3)} = \sqrt{(9)} = 3$, $\sqrt{(x + 10)} = 4$;
$x = 26$, $\sqrt{(2x - 3)} = \sqrt{(49)} = 7$, $\sqrt{(x + 10)} = 6$.

Thus $x = 26$ is the only solution.

Note. $x = 6$ is a solution of $\sqrt{(x + 10)} - \sqrt{(2x - 3)} = 1$.

EXERCISE 1.1

1. Express in terms of $\sqrt{2}$:
(i) $\sqrt{8}$, (ii) $\sqrt{18}$, (iii) $\sqrt{50}$, (iv) $\sqrt{(24 \cdot 5)}$.

2. Express in the form $a + b\sqrt{3}$:
(i) $(2 + \sqrt{3})(\sqrt{3} - 1)$, (ii) $(2 + \sqrt{3})/(\sqrt{3} - 1)$,
(iii) $(\sqrt{5} + \sqrt{3})/(\sqrt{5} - \sqrt{3})$, (iv) $(\sqrt{2} - 2\sqrt{3})/(2\sqrt{2} + \sqrt{3})$.

3. Starting with $x = 1$, apply Newton's formula $y = \frac{1}{2}(x + 3/x)$ three times and give the three successive approximations for $\sqrt{3}$.

Solve the equations in nos. 4–6:

4. $2\sqrt{x} + \sqrt{(2x + 1)} = 7$.

5. $\sqrt{(4x + 1)} - 2\sqrt{(2x)} + 1 = 0$.

6. $x\sqrt{(2 + x)} + 2\sqrt{(2 - x)} = \sqrt{(8 + x^3)}$.

7. Use the binary scale to find all the different weights in kilograms that could be measured with standard weights of 0·5, 1, 2, 4, 8, 16 kg.

8. Express the following numbers (i) in the scale of 3, and (ii) in the scale of 12:

25, 145, 356, 829.

9. For the quinary system, base 5, work out multiplication tables up to $4 \times 4 = \ldots$, and use these to express in the quinary scale:
(i) 341×24; (ii) $22033/42$.

Binary operations

As the science of arithmetic was developed and separated from the practical affairs of life, four basic operations were recognized; these operations are the formation of the sum, product, difference and quotient of two numbers.

If we form the sum or product of two *natural numbers* (the numbers derived from counting), we get another natural number, but the formation of the difference or quotient may lead to a number of a different type. We say that *the system of natural numbers is closed under the operations of addition and multiplication.* An operation which combines two numbers of a closed system and produces a number of the same system is called a *binary operation.*

In order to produce a system of numbers which is closed under subtraction it is necessary to extend the system to include all the *integers*, positive or negative, and zero; we thus obtain the system:

$$\ldots, -3, -2, -1, 0, 1, 2, 3, \ldots$$

In order to produce a system which is also closed under division (division by zero is excluded) we must further extend the number system to include all the *rational numbers*; examples of rational numbers are

$$\ldots, \quad -\tfrac{3}{2}, \quad -\tfrac{1}{1}, \quad 0, \quad \tfrac{1}{2}, \quad \tfrac{2}{3}, \quad \ldots$$

It will be seen that the rational numbers are not single numbers but number pairs. The operations of addition and multiplication for the rational numbers a/b and c/d are:

$$\frac{a}{b}+\frac{c}{d}=\frac{ad+bc}{bd} \qquad \frac{a}{d}\times\frac{c}{b}=\frac{ac}{bd}.$$

Quite apart from practical needs such as measuring, the extension of number systems has taken place through the desire of the mathematicians to find solutions of a variety of equations. Clearly this is a later development in the history of the subject, since algebraic equations could not be handled efficiently until an adequate notation was developed in the sixteenth century. Very briefly we can associate the natural numbers, integers, rational and real numbers with equations of the following types:

$$x-a=0, \quad x+a=0, \quad ax+b=0, \quad ax^2=b^2$$

where a is a positive integer and b is an integer.

The idea of number will be further extended in Chapter 8 to include solutions of all quadratic equations $ax^2+bx+c=0$, but we shall now review as simply as possible the stage that we have already reached. We think of a closed system of numbers as being defined rather by the properties of the binary operations that can be performed, than by the properties of the numbers themselves.

Axioms

Suppose x, y and z represent real numbers. When a letter is used in this way, as a place-holder which can be replaced by any number from a given system, it is called a *variable*.

The properties of the operations of addition and multiplication when applied to real numbers are as follows:

1. The closure law

$$x + y \text{ and } xy \text{ are real numbers.}$$

2. The commutative law

$$x + y = y + x, \quad xy = yx.$$

3. The associative law

$$\begin{aligned} x + (y + z) &= (x + y) + z, & x(yz) &= (xy)z, \\ &= x + y + z; & &= xyz. \end{aligned}$$

4. The law of identity

Numbers 0 and 1 exist such that

$$x + 0 = 0 + x = x, \quad x \times 1 = 1 \times x = x.$$

5. The law of the inverse

Corresponding to x there exist real numbers $(-x)$ and (x^{-1}) such that

$$x + (-x) = (-x) + x = 0, \; x \times (x^{-1}) = (x^{-1}) \times x = 1, \; (x \neq 0).$$

6. The distributive law

$$x(y + z) = xy + xz, \quad (y + z)\,x = yx + zx.$$

In deducing consequences of these laws, or postulates, we shall make use of rules of logic which are often applied more or less unconsciously. Among these rules are three that are of almost universal application:

Reflexive law: $x = x$,
Symmetric law: $x = y \Leftrightarrow y = x$,
Transitive law: $x = y$ and $y = z \Rightarrow x = z$.

These three laws clarify the meaning of equality and greatly assist the use of symbols.

There is one further law relating to the binary operations of addition and multiplication which must be clearly stated; it has such wide application that it ranks as a logical axiom:

Uniqueness: (i) $a = b$ and $c = d \Rightarrow a + c = b + d$,
(ii) $a = b$ and $c = d \Rightarrow ac = bd$.

We shall now consider a few of the useful results, already well known to the reader, which can be deduced from the axioms and postulates considered above.

Deductive methods

Rule 5 (Inverse) allows us to define subtraction and division in terms of addition and multiplication, so that subtraction and division are inverse operations:

$$a - b = a + (-b), \quad a/b = a \times (b^{-1}).$$

Cancellation (i) $a + b = a + c \Rightarrow b = c$;
(ii) $ab = ac \Rightarrow b = c \; (a \neq 0)$.

(i) $a + b = a + c$, (Given)
$(-a) + (a + b) = (-a) + (a + c)$, (Uniqueness)
$(-a + a) + b = (-a + a) + c$, (Associative)
$0 + b = 0 + c$, (Inverse)
$b = c$. (Identity)

(ii) $ab = ac$, (Given)
$(a^{-1})\, ab = (a^{-1})\, ac$, (Uniqueness)
$(a^{-1}\, a)\, b = (a^{-1}\, a)\, c$, (Associative)
$1 \times b = 1 \times c$, (Inverse, $a \neq 0$)
$b = c$. (Identity)

EXAMPLE. *Prove that (i)* $a \times 0 = 0$, *(ii)* $(-a)(-b) = ab$, *(iii)* $(a)(-b) = (-a)(b) = -ab$.

(i) $a + 0 = a$, (Identity)
$a(a + 0) = a^2$, (Uniqueness)
$a^2 + a \times 0 = a^2$, (Distributive)
$a^2 + a \times 0 = a^2 + 0$, (Identity)
$a \times 0 = 0$. (Cancellation)

(ii) $ab + [a(-b) + (-a)(-b)] = [ab + a(-b)] + (-a)(-b)$, (Associative)
$ab + [a(-b) + (-a)(-b)] = ab + [a + (-a)]\,(-b)$, (Distributive)
$= ab + 0 \times (-b)$, (Inverse)
$= ab + (-b) \times 0$, (Commutative)
$= ab + 0$, (Part (i))
$= ab$. (Identity)

Also

$$[ab + a(-b)] + (-a)(-b) = a[b + (-b)] + (-a)(-b), \quad \text{(Distributive)}$$
$$= a \times 0 + (-a)(-b), \quad \text{(Inverse)}$$
$$= 0 + (-a)(-b), \quad \text{(Part (i))}$$
$$= (-a)(-b). \quad \text{(Identity)}$$

Hence $$(-a)(-b) = ab. \quad \text{(Transitive)}$$

(iii) The reader should prove this in a manner similar to that followed in (ii); it is stated here because it will be used explicitly in the next section, on *order*. Meanwhile it should be noted that the law of closure was applied implicitly at several points.

The six laws, together with the logical axioms, expose some of the most important properties of the binary operations with which we are already familiar. In subsequent chapters we shall consider (i) *the algebra of sets*, one of the most fundamental of all mathematical studies, (ii) *the algebra of vectors*, an example of a double system of scalars and vectors, (iii) *the arithmetic of congruences*, where the number field is finite and where cancellation is not always possible, and (iv) *the algebra of complex numbers*.

In Volume 2 the study of vectors and of complex numbers will be taken further, and the algebra of matrices will be introduced. But first we shall consider further the special properties of real numbers, the properties of order.

Order

Any set of elements for which two binary operations are defined which satisfy the six laws on p. 12, is called a *field*. The real numbers, which we have related in a fundamental way to points on a line, form an *ordered field*. The intuitive ideas of order derived from the consideration of the points on a line are not a satisfactory basis for logical development of ideas, and so adequate postulates must be formulated.

1. $a < b \Leftrightarrow b - a > 0$.
2. One, and only one, of $a < b$, $a = b$, $b < a$ is always true.
3. $a > 0$ *and* $b > 0 \Rightarrow a + b > 0$.
4. $a > 0$ *and* $b > 0 \Rightarrow ab > 0$.
5. If $a < b$, there always exists c, such that $a < c < b$.
6. If a is given, there always exist b and c such that $b < a < c$.

EXAMPLE. *If $a < b$, prove that (i) $a + c < b + c$ for any c; (ii) $ac < bc$ when $c > 0$; (iii) $ac > bc$ when $c < 0$; (iv) $a < b$ and $b < c \Rightarrow a < c$.*

(i)
$$a < b \Leftrightarrow b - a > 0.$$
But
$$(b + c) - (a + c) = b - a \text{ for any } c,$$
and so
$$(b + c) - (a + c) > 0,$$
or
$$a + c < b + c.$$

(ii)
$$a < b \Leftrightarrow b - a > 0.$$
If also
$$c > 0, \text{ then } (b - a)\, c > 0,$$
so that
$$bc - ac > 0,$$
or
$$ac < bc.$$

(iii)
$$\begin{gathered} \text{If } a < b \text{ and } c < 0, \text{ then} \\ b - a > 0 \text{ and } (-c) > 0, \\ (b - a)(-c) > 0 \\ ac - bc > 0 \end{gathered}$$
and so
$$ac > bc.$$

(iv)
$$\begin{gathered} a < b \text{ and } b < c \text{ gives} \\ (b - a) > 0 \text{ and } (c - b) > 0, \end{gathered}$$
so that
$$(b - a) + (c - b) > 0,$$

or

$$c - a > 0 \Leftrightarrow a < c.$$

Note. The four rules of the previous Example form the basis of much work on inequalities, and should be well known.

EXAMPLE. *Find the possible range of values of x when*

$$-1 < \frac{x+1}{3x-1} < 1.$$

Method (i). When $3x - 1 > 0$, multiplication of the inequality by $(3x - 1)$ gives

$$-3x + 1 < x + 1 < 3x - 1,$$

so that

$$-3x + 1 < x + 1 \Rightarrow x > 0$$

and

$$x + 1 < 3x - 1 \Rightarrow x > 1.$$

Also $3x - 1 > 0 \Rightarrow x > \frac{1}{3}$, and the strictest of the three conditions is $x > 1$.

When $3x - 1 < 0$, multiplication of the inequality by $(3x - 1)$ gives

$$-3x + 1 > x + 1 > 3x - 1,$$

so that

$$-3x + 1 > x + 1 \Rightarrow x < 0,$$

and

$$x + 1 > 3x - 1 \Rightarrow x < 1,$$

while

$$3x - 1 < 0 \Rightarrow x < \tfrac{1}{3}, \text{ and so } x < 0.$$

The inequality is therefore satisfied either when $x < 0$ or when $x > 1$.

Method (ii). Let

$$a = \frac{x+1}{3x-1} + 1 = \frac{4x}{3x-1}.$$

Thus

$$a = 4x(3x - 1)/(3x - 1)^2, \text{ and } (3x - 1)^2 > 0,$$

so that

$$a > 0 \text{ when } x < 0 \text{ or when } x > \tfrac{1}{3}.$$

Let

$$b = 1 - \frac{x+1}{3x-1} = \frac{2(x-1)}{3x-1}.$$

Then

$$b = 2(x-1)(3x-1)/(3x-1)^2,$$

and so

$$b > 0 \text{ when } x > 1 \text{ or when } x < \tfrac{1}{3}.$$

Both

$$a > 0 \text{ and } b > 0 \text{ when } x < 0 \text{ or when } x > 1.$$

The reader will probably feel that Method (ii) contains one or two steps which go rather beyond the stated postulates and the properties proved in the Example above, but it presents ideas which are very useful in dealing with inequalities.

EXERCISE 1.2

Prove the properties of inequalities in nos. 1–5*:*

1. $a < b$ *and* $c < d \Rightarrow a + c < b + d$.

2. $a < b$ *and* $ab > 0 \Rightarrow 1/a > 1/b$.

3. $ab > 0 \Rightarrow$ *either* $a > 0, b > 0$ *or* $a < 0, b < 0$.

4. $a > 0 \Rightarrow 1/a > 0$; $a < 0 \Rightarrow 1/a < 0$.

5. $a > b > 0 \Rightarrow 1/b > 1/a > 0$.

In nos. 6–14 *find the possible range of values of x:*

6. $x + 3 > 0$ and $x - 2 < 0$.

7. $2x + 3 > 0$, and $2x - 1 < 0$.

8. $(x + 2)(x - 1) < 0$.

9. $x^2 > 9$.

10. $x^2 < 25$.

11. $1/(x - 1) > 2/(x - 3)$.

12. $\dfrac{2}{2x-1} > \dfrac{3}{3x-1}$.

13. $-1 < \dfrac{x}{2x-1} < 1$.

14. $-1 < \dfrac{4x-1}{x+5} < 1$.

15. Give integral values of x such that
$$1 - 2\sqrt{3} < x < 1 + 2\sqrt{3}.$$

16. What is the least value of
(i) $(x - 1)^2 + 3$, (ii) $x^2 + 4x + 3$?

17. What is the greatest value of
(i) $5 - (2x + 1)^2$, (ii) $3 + 2x - x^2$?

18. Find positive integers x, y such that
(i) $x < 4, \quad xy < 15$;
(ii) $x + y \leqslant 6, \quad xy \geqslant 9$;
(iii) $x + y \leqslant 4, \quad xy \geqslant 3$.

19. Give an example of integers a, b, c, d to show that $a > b, c > d$ can imply that $ac < bd$.

CHAPTER 2

SETS

Definition of a set

The theory of sets originated with the German mathematician G. Cantor (1845–1918), who defined a set **S** as a collection of definite, distinguishable objects which are selected by means of some rule or description. The objects are called the *elements* or members of **S**, but the set exists only when the collection is treated as a whole.

If the set **S** is composed of the digits 0, 1, 2, 3 we write

$$\mathbf{S} = \{0, 1, 2, 3\},$$

and this is an example of a *finite set*, defined by the actual display of the elements.

If the set **S** is composed of all the positive even numbers we can write

$$\mathbf{S} = \{a \mid a \text{ is a positive even number}\},$$

which is simply an abbreviation for "**S** is the set of all numbers a such that a is a positive even number". The vertical line $|$ is used for "such that". This is an example of an *infinite set*. If it is known that n is a positive integer, the same set can be described by

$$\mathbf{S} = \{a \mid a = 2n\}.$$

Membership of a set is denoted by the symbol $\in$, and we write

$$x \in \mathbf{S}$$

as an abbreviation for "x is an element of the set **S**". The contradiction "x is NOT an element of **S**" is written

$$x \notin \mathbf{S}.$$

We use $x, y, z \in \mathbf{S}$ to indicate that $x \in \mathbf{S}$ *and* $y \in \mathbf{S}$ *and* $z \in \mathbf{S}$.

Sets are not necessarily connected with numbers; any collection of objects or persons which are definite and distinguishable can form a set if they are thought of as a whole.

Equal sets

Two sets **P** and **Q** are equal if and only if they consist of the same elements; in other words **P** = **Q** if and only if every element p of **P** is also an element of **Q**, and if every element q of **Q** is also an element of **P**. Thus **P** = **Q** means that

$$p \in \mathbf{P} \Leftrightarrow p \in \mathbf{Q}.$$

Ex. 1. Are the following sets equal?
(i) $\{2, 4, 6\}$ and $\{4, 6, 2\}$, (ii) $\{2, 4, 6\}$ and $\{2, 4, 6, 6\}$.

Ex. 2. What can be deduced from $\{a, b\} = \{c, d\}$?

Subsets

Other sets can be formed from a given set by removing some of its elements. Thus, if

$$\mathbf{S} = \{a, b, c, d\}$$

we can form *subsets* $\{a, b, d\}$, $\{b, d\}$, $\{a\}$ and so on. It is convenient to regard both the set **S**, formed by removing no members, and the *empty set*, formed by removing all the members, as subsets of **S**. The set **S** = $\{a, b, c, d\}$ therefore contains the sixteen subsets

$\{a, b, c, d\}$	$\{b, c, d\}$	$\{a, c, d\}$	$\{a, b, d\}$
$\{a, b, c\}$	$\{a, b\}$	$\{a, c\}$	$\{a, d\}$
$\{b, c\}$	$\{b, d\}$	$\{c, d\}$	$\{a\}$
$\{b\}$	$\{c\}$	$\{d\}$	$\{\ \}$.

The empty set, $\{\ \}$, is sometimes denoted by Φ.

The fact that the set **P** is a subset of **S** is written $\mathbf{P} \subseteq \mathbf{S}$, or $\mathbf{S} \supseteq \mathbf{P}$. Thus $\mathbf{P} \subseteq \mathbf{S}$ means that

$$p \in \mathbf{P} \Rightarrow p \in \mathbf{S}.$$

$\mathbf{P} \subset \mathbf{S}$ means that **P** is a subset of **S** but is not equal to **S**; in this case **P** is a *proper subset* of **S**, or **P** is *properly included* in **S**.

The fact that **P** is not a subset of **S** is written **P** $\nsubseteq$ **S**.

EXAMPLE. *Prove that, if $A \subseteq B$ and $B \subseteq C$, then $A \subseteq C$.*

If $A \subseteq B$, then $x \in A \Rightarrow x \in B$;

if $B \subseteq C$, then $x \in B \Rightarrow x \in C$.

Hence $x \in A \Rightarrow x \in C$, and so

$$A \subseteq C.$$

Ex. 3. How many subsets has **S** when the number of its elements is (i) 2, (ii) 3, (iii) n?

Ex. 4. Prove that $X \subseteq Y$ and $Y \subseteq X \Leftrightarrow X = Y$.

Operations on sets

If all the sets under consideration are subsets of a set $\mathscr{E}$, then $\mathscr{E}$ is called the *universal set.*

If the rectangle in Fig. 2.1 represents $\mathscr{E}$, while the circle represents a subset **A**, then the subset of $\mathscr{E}$ that is left when **A** is removed is called the *complement* of **A** and is written **A**′, or $\overline{\mathbf{A}}$; it is represented in Fig. 2.1 by the shaded area.

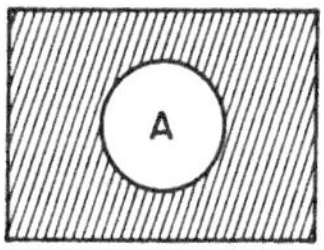

Fig. 2.1

In Fig. 2.2 the circles represent the sets **A** and **B**. The set which is composed of all elements which are members of **A** or of **B**, or of both **A** and **B**, is called the *union* of **A** and **B**; it is written **A** ∪ **B** and read "**A** cup **B**". In Fig. 2.2, **A** ∪ **B** is represented by the shaded area.

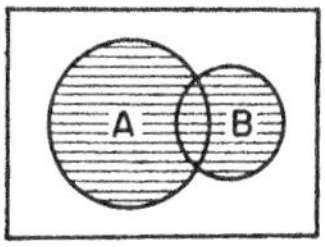

Fig. 2.2

The set which is composed of all elements which are members both of **A** and of **B** is called the *intersection* of **A** and **B**; it is written $\mathbf{A} \cap \mathbf{B}$ and read "**A** cap **B**". In Fig. 2.3, $\mathbf{A} \cap \mathbf{B}$ is represented by the shaded area.

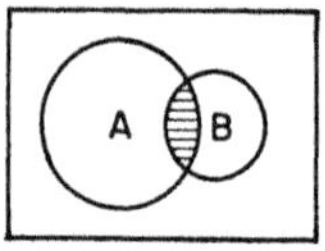

Fig. 2.3

Two sets **A** and **B** are said to be *disjoint* if they have no common member; thus **A** and **B** are disjoint if and only if

$$\mathbf{A} \cap \mathbf{B} = \Phi.$$

The graphic representation of sets, as in Figs. 2.1 to 2.3, is called a *Venn diagram*. Such diagrams may be used in the intuitive consideration of the properties of the relations of complementation, union and intersection of sets. In a strict axiomatic theory this intuitive consideration must be replaced by deduction from definitions by means of axioms or rules, as illustrated in the Example on p. 23.

The algebra of sets

The binary operations of union and intersection possess properties very similar to those of addition and multiplication in the algebra of numbers.

1. **Closure**

$X \cup Y$ and $X \cap Y \subseteq \mathscr{E}$.

2. **Commutative**

$$X \cup Y = Y \cup X, \qquad X \cap Y = Y \cap X.$$

3. **Associative**

$$X \cup (Y \cup Z) = (X \cup Y) \cup Z,$$
$$X \cap (Y \cap Z) = (X \cap Y) \cap Z.$$

4. **Identity**

$$X \cup \Phi = \Phi \cup X = X, \qquad X \cap \mathscr{E} = \mathscr{E} \cap X = X.$$

5. **Inverse**

$$X \cup X' = X' \cup X = \mathscr{E}, \qquad X \cap X' = X' \cap X = \Phi.$$
$$X \cup \mathscr{E} = \mathscr{E} \cup X = \mathscr{E}, \qquad X \cap \Phi = \Phi \cap X = \Phi.$$

6. **Distributive**

$$X \cap (Y \cup Z) = (X \cap Y) \cup (X \cap Z),$$
$$X \cup (Y \cap Z) = (X \cup Y) \cap (X \cup Z).$$

An organized system of sets subject to two binary operations which follow these laws is called a *Boolean algebra*, after George Boole (1815–1864).

Ex. 5. Verify law 6 above for the sets X, Y, Z when
(i) $X = \{1, 2\}$, $Y = \{1, 2, 3\}$, $Z = \{2, 3, 4\}$;
(ii) $X = \{a, b, d\}$, $Y = \{d, c, a, e\}$, $Z = \{b, e, d\}$.

Ex. 6. Illustrate laws 1–6 above in Venn diagrams.

Ex. 7. $A \cup B = \{x | x \in A \text{ or } x \in B\}$; give a similar definition for $A \cap B$.

Ex. 8. When is it true that (i) $A \cup B = A \cap B$, (ii) $A \cap B = B$, (iii) $A \cap B' = A$, (iv) $A \cap B' = \Phi$?

EXAMPLE. *Prove that* $X \cap (Y \cup Z) = (X \cap Y) \cup (X \cap Z)$.

$$\begin{aligned} Y \cup Z &= \{x | x \in Y \text{ or } x \in Z\}. \\ X \cap (Y \cup Z) &= \{x | x \in X \text{ and } (x \in Y \text{ or } x \in Z)\} \\ &= \{x | x \in X \text{ and } x \in Y\} \text{ or } \{x | x \in X \text{ and } x \in Z\} \\ &= (X \cap Y) \cup (X \cap Z). \end{aligned}$$

Union and intersection tables

Consider the set $\mathscr{E} = \{x, y, z\}$ and its eight subsets:

$$\mathscr{E} = \{x, y, z\}, \quad \mathbf{A} = \{y, z\}, \quad \mathbf{B} = \{x, z\}, \quad \mathbf{C} = \{x, y\}$$
$$\mathbf{P} = \{x\}, \quad \mathbf{Q} = \{y\}, \quad \mathbf{R} = \{z\}, \quad \Phi = \{\ \}.$$

A table of unions for the eight subsets on the lines of an addition table gives the following results:

$\cup$	$\mathscr{E}$	A	B	C	P	Q	R	Φ
$\mathscr{E}$	$\mathscr{E}$	$\mathscr{E}$	$\mathscr{E}$	$\mathscr{E}$	$\mathscr{E}$	$\mathscr{E}$	$\mathscr{E}$	$\mathscr{E}$
A	$\mathscr{E}$	A	$\mathscr{E}$	$\mathscr{E}$	$\mathscr{E}$	A	A	A
B	$\mathscr{E}$	$\mathscr{E}$	B	$\mathscr{E}$	B	$\mathscr{E}$	B	B
C	$\mathscr{E}$	$\mathscr{E}$	$\mathscr{E}$	C	C	C	$\mathscr{E}$	C
P	$\mathscr{E}$	$\mathscr{E}$	B	C	P	C	B	P
Q	$\mathscr{E}$	A	$\mathscr{E}$	C	C	Q	A	Q
R	$\mathscr{E}$	A	B	$\mathscr{E}$	B	A	R	R
Φ	$\mathscr{E}$	A	B	C	P	Q	R	Φ

Ex. 9. (i) Which subsets are the complements of **A**, **P** ?
(ii) Verify that $\mathbf{A} \cup \mathbf{A}' = \mathscr{E}$, $\quad (\mathbf{A}')' = \mathbf{A}$.

A table of intersections for the same eight subsets on the lines of a multiplication table gives the following results:

$\cap$	$\mathscr{E}$	A	B	C	P	Q	R	Φ
$\mathscr{E}$	$\mathscr{E}$	A	B	C	P	Q	R	Φ
A	A	A	R	Q	Φ	Q	R	Φ
B	B	R	B	P	P	Φ	R	Φ
C	C	Q	P	C	P	Q	Φ	Φ
P	P	Φ	P	P	P	Φ	Φ	Φ
Q	Q	Q	Φ	Q	Φ	Q	Φ	Φ
R	R	R	R	Φ	Φ	Φ	R	Φ
Φ	Φ	Φ	Φ	Φ	Φ	Φ	Φ	Φ

Ex. 10. Verify that $\mathbf{A} \cap \mathbf{A}' = \Phi$, $\quad \mathbf{P} \cap \mathbf{P}' = \Phi$.

Ex. 11. Verify that (i) $(\mathbf{A} \cup \mathbf{B})' = \mathbf{A}' \cap \mathbf{B}'$, (ii) $(\mathbf{A} \cap \mathbf{B})' = \mathbf{A}' \cup \mathbf{B}'$.

EXAMPLE. *If $A \cup X = \mathscr{E}$ and $A \cap X = \Phi$, prove that $X = A'$.*

$$\begin{aligned} X \cap (A \cup A') &= (X \cap A) \cup (X \cap A') \\ &= \Phi \cup (X \cap A') \\ &= X \cap A'. \\ X \cap (A \cup A') &= X \cap \mathscr{E} = X. \end{aligned}$$

Hence

$$X \cap A' = X. \tag{i}$$

Since

$$A \cup X = \mathscr{E},$$
$$A' \cap (A \cup X) = A' \cap \mathscr{E}$$
$$= A',$$

and so

$$(A' \cap A) \cup (A' \cap X) = A',$$
$$\Phi \cup (A' \cap X) = A'.$$

Hence

$$A' \cap X = A'. \qquad \text{(ii)}$$

From (i) and (ii),

$$X = A'.$$

EXAMPLE. *Prove that (i)* $(A \cup B)' = A' \cap B'$,
(ii) $(A \cap B)' = A' \cup B'$.

(i)
$$\begin{aligned}(A \cup B) \cap (A' \cap B') &= [(A' \cap B') \cap A] \cup [(A' \cap B') \cap B] \\ &= [(A \cap A') \cap B'] \cup [(B \cap B') \cap A'] \\ &= (\Phi \cap B') \cup (\Phi \cap A') \\ &= \Phi.\end{aligned}$$
$$\begin{aligned}(A \cup B) \cup (A' \cap B') &= [(A \cup B) \cup A'] \cap [(A \cup B) \cup B'] \\ &= [(A \cup A') \cup B] \cap [A \cup (B \cup B')] \\ &= (\mathscr{E} \cup B) \cap (A \cup \mathscr{E}) \\ &= \mathscr{E} \cap \mathscr{E} \\ &= \mathscr{E}.\end{aligned}$$

Using the result of the previous Example,

$$A' \cap B' = (A \cup B)'.$$

(ii) Interchanging all subsets and their complements, we have

$$(A \cap B)' = A' \cup B'.$$

De Morgan's rule

The results of the previous Example are

$$\mathbf{(A \cup B)' = A' \cap B'}, \qquad \mathbf{(A \cap B)' = A' \cup B'}.$$

Expressed in words, these become:

The complement of a union is the intersection of the complements, and the complement of an intersection is the union of the complements.

Boolean algebras have important applications in the design of electrical switching circuits and in electronic computers. They are also applicable, with a slight change of notation, to symbolic logic, which was Boole's main interest. In symbolic logic a letter represents not a set but a statement or proposition. The symbol $\cup$ is used for "or" in the sense of "either one, or the other, or both", and $\cap$ is used for "and" in the sense of "both one and the other at the same time"; the complement of a set **A** becomes the contradiction of the statement **A**, or "not-**A**"; $\mathbf{A} = \mathscr{E}$ means that **A** is true; $\mathbf{A} = \Phi$ means that **A** is false. With these modifications a class of statements and their logical combinations becomes a Boolean algebra.

$\mathbf{A} - \mathbf{B}$

If **A** and **B** are any sets, $\mathbf{A} - \mathbf{B}$ denotes the set of elements of **A** which are not elements of **B**; in other words $\mathbf{A} - \mathbf{B} = \mathbf{A} \cap \mathbf{B}'$.

Note that $\mathbf{A} - \mathbf{A} = \mathbf{A} \cap \mathbf{A}' = \Phi$, $\mathbf{A} - \Phi = \mathbf{A} \cap \mathscr{E} = \mathbf{A}$.

Ex. 12. Prove that (i) $\mathbf{A} \cap (\mathbf{B} - \mathbf{C}) = \mathbf{A} \cap \mathbf{B} - \mathbf{C}$,
(ii) $\mathbf{A} - \mathbf{B} = \mathbf{A} - \mathbf{A} \cap \mathbf{B}$.

EXERCISE 2.1

1. If $\mathscr{E} = \{a, b, c, d, e, f, g, h\}$, $X = \{a, b, c, d\}$, $Y = \{a, b, e, f, g\}$, write down the sets
(i) $X \cup Y$, (ii) $X \cap Y$, (iii) $X' \cup Y'$, (iv) $X' \cap Y'$, (v) $(X \cup Y)'$, (vi) $(X \cap Y)'$, (vii) $(X' \cup Y')'$, (viii) $(X' \cap Y')'$.
Do these results verify de Morgan's rule?

2. If P, Q, R are subsets of $\mathscr{E}$, illustrate the following by Venn diagrams:
(i) $P \cap Q$, (ii) $P' \cup R$, (iii) $Q' \cap R'$, (iv) $Q' \cup R'$, (v) $P \cup Q \cup R$, (vi) $P \cap Q \cap R$, (vii) $(P \cap Q) \cup R$, (viii) $P' \cap Q' \cap R'$, (ix) $P \subseteq Q$, (x) $(P \cup Q) \subseteq R$, (xi) $(P \cap Q) \subseteq R$, (xii) $(P \cap Q) \subseteq (Q \cap R)$.

3. Prove formally that $A \cup (B \cap C) = (A \cup B) \cap (A \cup C)$.

4. Give simpler alternative forms for the following:
(i) $(P \cup Q)'$, (ii) $(P' \cup Q')'$, (iii) $(P \cup Q) \cap (Q \cup R)$,
(iv) $(P \cap Q') \cup (P \cap R)'$, (v) $(P \cap Q)' \cap (P \cap R)'$,
(vi) $P \cap (P' \cap Q)$, (vii) $[(A \cap B') \cup (B \cap C')] \cap (A \cap C')$.

5. Show by an example that "subtraction" of sets as defined above is not (i) commutative, (ii) associative.

6. Prove the following:
(i) $A \cap (B - C) = (A \cap B) - (A \cap C)$, (ii) $(A - B)' = B \cup A'$, (iii) $A - (A - B) = A \cap B$, (iv) $(A - B) \cap (A - C) = A - (B \cup C)$, (v) $(A - C) \cup (B - C) = (A \cup B) - C$, (vi) $(A - C) \cap (B - C) = (A \cap B) - C$, (vii) $(A - B) \cap (A - C) = (A - B) - C$, (viii) $(A - B) \cup (A \cap B \cap C) = A - (B - C)$.

Counting sets

The fundamental idea from which counting derives is that of the *one–one correspondence*. If the elements of a set **A** can be exactly matched with the elements of a set **B**, each element of **A** being paired with a unique element of **B** so that no element of either set is left unpaired, we say there is a one–one correspondence between the sets. For example, if there are six married couples in a room, there is a one–one correspondence between the set of men and the set of women.

When two sets are in one–one correspondence they are said to be equi-numerous, or to have the same *cardinal number*. A set is said to be *countable* if its elements can be placed in one–one correspondence with the positive integers. If the set is infinite but a one–one correspondence with the positive integers can be established, the set is said to be *countably infinite*; for example, the set of even integers is countably infinite since the even integer $2n$ corresponds to the integer n.

The cardinal number of a finite set **A** is denoted by $N(\mathbf{A})$. If **A** and **B** are disjoint sets

$$N(\mathbf{A} \cup \mathbf{B}) = N(\mathbf{A}) + N(\mathbf{B})$$

since, in finding $N(\mathbf{A} \cup \mathbf{B})$, each element of **A** and each element of **B** is counted once and once only.

When two sets are not disjoint, like the sets **A** and **B** in Fig. 2.4, we can see that the elements in $\mathbf{A} \cap \mathbf{B}$ will be counted twice when we evaluate $N(\mathbf{A}) + N(\mathbf{B})$; it appears that

$$N(\mathbf{A} \cup \mathbf{B}) = N(\mathbf{A}) + N(\mathbf{B}) - N(\mathbf{A} \cap \mathbf{B}).$$

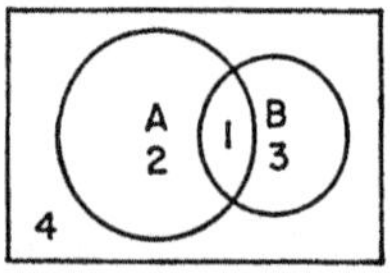

Fig. 2.4

We can prove this deductively by breaking down the sets into disjoint subsets; for example, the set **A** can be divided into the disjoint subsets $\mathbf{A} \cap \mathbf{B}'$ and $\mathbf{A} \cap \mathbf{B}$. (These subsets are represented by the areas marked 2 and 1 respectively in Fig. 2.4.) Similarly $\mathbf{B} = (\mathbf{A}' \cap \mathbf{B}) \cup (\mathbf{A} \cap \mathbf{B})$.

Thus

$$N(\mathbf{A}) = N(\mathbf{A} \cap \mathbf{B}') + N(\mathbf{A} \cap \mathbf{B}),$$
$$N(\mathbf{B}) = N(\mathbf{A}' \cap \mathbf{B}) + N(\mathbf{A} \cap \mathbf{B}).$$

Hence

$$N(\mathbf{A}) + N(\mathbf{B}) = N(\mathbf{A} \cap \mathbf{B}') + N(\mathbf{A}' \cap \mathbf{B}) + 2N(\mathbf{A} \cap \mathbf{B}).$$

But the sets $A \cap B'$, $A' \cap B$, $A \cap B$ are disjoint subsets whose union is $\mathbf{A} \cup \mathbf{B}$, and so

$$N(\mathbf{A} \cup \mathbf{B}) = N(\mathbf{A}) + N(\mathbf{B}) - N(\mathbf{A} \cap \mathbf{B}).$$

A similar result can be established for three sets **A**, **B** and **C**. It is

$$N(\mathbf{A} \cup \mathbf{B} \cup \mathbf{C}) = N(\mathbf{A}) + N(\mathbf{B}) + N(\mathbf{C})$$
$$-N(\mathbf{B} \cap \mathbf{C}) - N(\mathbf{A} \cap \mathbf{C}) - N(\mathbf{A} \cap \mathbf{B}) + N(\mathbf{A} \cap \mathbf{B} \cap \mathbf{C}).$$

In examining the results of statistical estimates of overlapping sets, we have as a check the fact that the cardinal number of a subset must be positive.

Ex. 13. Prove the result given above for $N(\mathbf{A} \cup \mathbf{B} \cup \mathbf{C})$, (i) by drawing a Venn diagram, (ii) by applying the previous result for $N(\mathbf{A} \cup \mathbf{B})$.

EXAMPLE. *A penny, a sixpence and a shilling were tossed together* 100 *times. The numbers of heads observed were respectively* 65, 44 *and* 58. *The penny and the sixpence showed heads together* 35 *times, the sixpence and shilling together* 27 *times. What can be said about the number of times that all three coins showed heads, or tails, together?*

Let P, Q, R denote the sets of heads shown by the penny, sixpence and shilling, so that $N(P) = 65$, $N(Q) = 44$, $N(R) = 58$, $N(P \cap Q) = 35$, $N(Q \cap R) = 27$.

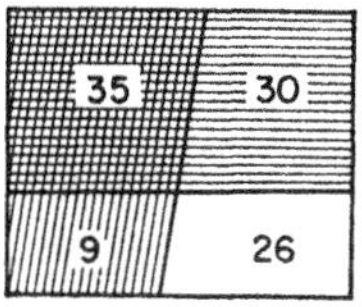

Fig. 2.5

Figure 2.5 shows P shaded horizontally and Q shaded obliquely, so that $P \cap Q$ is doubly shaded. But $N(Q \cap R) = 27$, so that, if $x = N(P \cap Q \cap R)$, $18 < x < 27$, since $Q \cap R$ can contain between 0 and 9 of P', the complement of P.

Also $N(R - Q) = 58 - 27 = 31$; it follows that the greatest value of $N(P' \cap Q' \cap R')$ is 25, while its least value is zero, since $P' \cap Q'$ (labelled 26) can be covered by $(R - Q)$ with 5 to spare.

A solution using the algebra of sets is more easily based on Fig. 2.6, where the numbers of certain subsets are marked x, a, b, c.

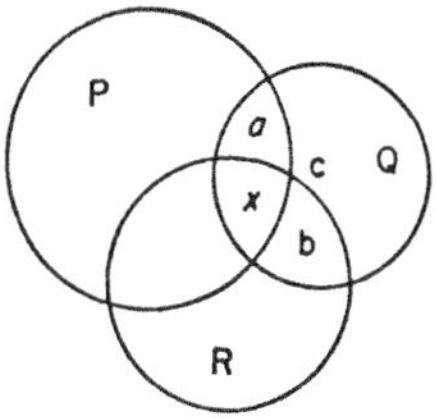

Fig. 2.6

We have

$$x + a + b + c = N(Q) = 44.$$

But

$$a = 35 - x, \quad b = 27 - x \Rightarrow 62 - x + c = 44,$$

and so

$$c = x - 18.$$

Since both b and c cannot be negative,

$$18 \leqslant x \leqslant 27. \qquad \text{(i)}$$

Again

$$N(P \cap Q)' = 100 - 35 \Rightarrow N(P' \cup Q') = 65,$$

and

$$N(P') = 35, \quad N(Q') = 56, \quad N(R') = 42.$$

But

$$N(P' \cup Q') = N(P') + N(Q') - N(P' \cap Q'),$$

and so

$$N(P' \cap Q') = 35 + 56 - 65 = 26.$$

But

$$P' \cap Q' \cap R' \subseteq P' \cap Q',$$

therefore

$$N(P' \cap Q' \cap R') \leqslant 26. \qquad \text{(ii)}$$

Similarly

$$N(Q \cap R)' = 100 - 27 \Rightarrow N(Q' \cup R') = 73,$$

and

$$N(Q' \cup R') = N(Q') + N(R') - N(Q' \cap R')$$
$$\Rightarrow N(Q' \cap R') = 56 + 42 - 73 = 25$$
$$\Rightarrow N(P' \cap Q' \cap R') \leqslant 25. \qquad \text{(iii)}$$

Condition (iii) is more strict than (ii). Hence, from (i) and (iii), the number of times when all three coins showed heads together was between 18 and 27, and they showed tails together not more than 25 times.

EXERCISE 2.2

1. The statement $P \subseteq Q$ can be written $P \cap Q = P$. Write it as an equation in two other ways.

2. When is the statement $(A \cup B) \cap B' = A$ true?

3. Prove that (i) $(A \cap B) \cup (A \cap B') = A$, (ii) $A \cup (A' \cap B) = A \cup B$.

4. In a class of 33 children, 26 have had mumps and 25 have had chicken-pox, what can you say (i) about the number who have had both, (ii) about the number who have had neither?

5. In no. 4 if 20 have had mumps and chicken-pox, and 15 have had measles, what can you say about the number who have had all three?

6. In a science class 12 students read physics, 12 read mathematics, 13 read chemistry, 5 read physics and chemistry, 3 read physics and mathematics, 6 read chemistry and mathematics, and 2 read all three subjects. How many students (i) read only mathematics, (ii) read chemistry but not mathematics, (iii) read physics and chemistry but not mathematics?

7. In a motor coach there are 9 boys, 9 British children, 15 men, 5 foreign boys, 23 British persons, 12 British males, and 7 foreign females. What is the number of people on the motor coach?

8. In a school of 550 boys, 290 studied French, 220 Latin and 80 no language. How many studied both French and Latin?

9. If, in no. 8, it is also given that 90 studied German and 50 studied both Latin and French, find how many studied German only. If also 23 studied both Latin and German, find how many studied French and German but not Latin. What can be said about those who studied all three languages?

10. Apply the algebra of sets to prove that

$$A \cap B = \Phi \Leftrightarrow A \subseteq B'.$$

11. Prove that $A = B = \Phi \Leftrightarrow A \cup B = \Phi$.
[$A \cup B = \Phi \Rightarrow A' \cap (A \cup B) = A' \cap \Phi$, and so on.]

12. In a school all boys play one or more of the games, cricket, tennis, football. Sixty per cent play cricket, seventy-five per cent play tennis and seventy per cent play football; forty-five per cent play cricket and tennis, forty per cent play cricket and football, and fifty per cent play tennis and football. What percentage of the boys plays all three games?

CHAPTER 3

VECTORS AND CONGRUENCES

Directed lengths

Figure 3.1 shows three views of a directed line AB on which is chosen an origin, O. The set of points on this line provides a representation of the set of real numbers, with zero at O, positive numbers $(0 < x)$ to the right of O and negative numbers $(x < 0)$ to the left of O.

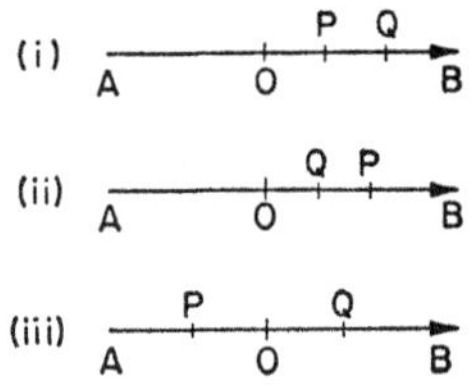

Fig. 3.1

P and Q are two points on AB, and the length of the directed segment $\overrightarrow{PQ}$ is a directed number d. If Q is to the right of P, d is positive; if Q is to the left of P, d is negative.

$\overrightarrow{OP}$ and $\overrightarrow{OQ}$ are directed numbers, p and q. We can see from Fig. 3.1 that, in all three cases,

$$\overrightarrow{OP} + \overrightarrow{PQ} = \overrightarrow{OQ},$$

or

$$p + d = q.$$

For example, in (iii),

$$\overrightarrow{PO} + \overrightarrow{OQ} = \overrightarrow{PQ},$$

and so

$$(-p)+q=d, \quad \text{or} \quad p+d=q.$$

Ex. 1. Figure 3.1 shows only three of six possible positions of P and Q in relation to O. Draw figures to show the remaining three possibilities, and show that, in all cases, $\overrightarrow{OP}+\overrightarrow{PQ}=\overrightarrow{OQ}$, and $p+d=q$.

Ex. 2. If X is the mid-point of PQ, so that $\overrightarrow{PX}=\overrightarrow{XQ}$, show that $\overrightarrow{OX}=\frac{1}{2}(\overrightarrow{OP}+\overrightarrow{OQ})$, or $x=\frac{1}{2}(p+q)$.

Ordered pairs

In general the order in which the elements of a set are displayed does not matter; for example,

$$\{a, b, c, d\}=\{b, c, a, d\}=\{d, b, a, c\}$$

and so on. There are, however, important cases where the order of the elements in a set does matter, where a certain element must be placed first, another element second. For example, $\{a, b\}$ is an *ordered pair* when a must be taken first and b second; we write it (a, b).

Two ordered pairs (a, b) and (c, d) are therefore equal when and only when $a=c$ *and* $b=d$. Also $(a, b)=(b, a)$ only when $a=b$.

Coordinates

Consider now the position of a point in a plane referred to an origin O and two directed axes OX, OY lying in the plane. OY is at right-angles to OX, as in Fig. 3.2.

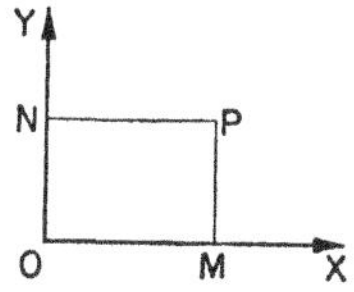

Fig. 3.2

P is a given point in the plane. PM, PN are the perpendiculars from P to OX, OY. The coordinates of P are the directed lengths

$\overrightarrow{OM} = x$ and $\overrightarrow{ON} = y$. We write P is the point (x, y), or, more briefly, P is (x, y). Here (x, y) is an ordered pair of real numbers.

The set of all ordered pairs (x, y) is in one–one correspondence with the set of points in the plane. If $\mathbf{X} = \{x\}$ and $\mathbf{Y} = \{y\}$, the set of ordered pairs (x, y) is called the cartesian product of $\mathbf{X}$ and $\mathbf{Y}$, and is written $\mathbf{X} \times \mathbf{Y}$.

Vectors

In Fig. 3.3 we can think of P as determining a displacement $\overrightarrow{OP}$. If QR is equal and parallel to OP, a translation of the plane

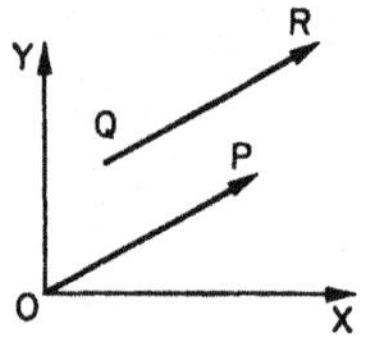

Fig. 3.3

which carried O to P would carry Q to R. We write

$$\overrightarrow{QR} = \overrightarrow{OP}.$$

A quantity which can be represented by a directed line such as $\overrightarrow{OP}$ or $\overrightarrow{QR}$ is called a *vector*. For example, displacements, velocities, forces are all vectors.

Figure 3.4 shows the method of combining vectors by the

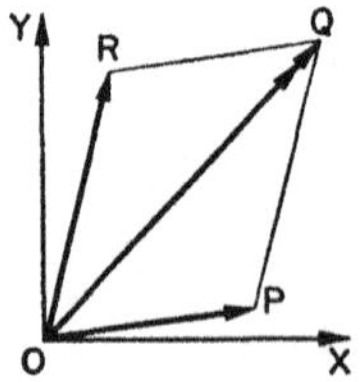

Fig. 3.4

parallelogram law. For example, a displacement $\overrightarrow{OP}$ followed by a displacement $\overrightarrow{OR}$ produces a resultant displacement $\overrightarrow{OQ}$, where $OPQR$ is a parallelogram. In the science laboratory, experiments are designed to show that the resultant of two forces is found in the same way.

Figure 3.5 shows that a triangle law is sufficient to determine

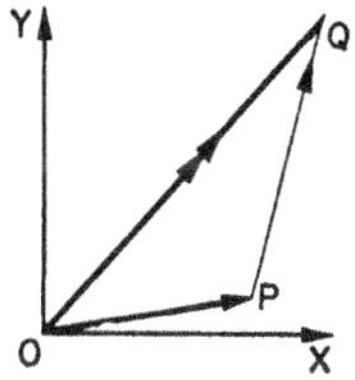

Fig. 3.5

the same vector $\overrightarrow{OQ}$. We write

$$\overrightarrow{OQ} = \overrightarrow{OP} + \overrightarrow{PQ},$$

the + sign being used to indicate the operation of combining two vectors by means of the parallelogram or triangle law. This operation is called the *addition of vectors*. Since the results of addition of two vectors is another vector, it is a binary operation.

It is customary to use a single letter to represent a vector. For example, in Fig. 3.5, if $\overrightarrow{OP} = \overrightarrow{p}$, $\overrightarrow{OR} = \overrightarrow{r}$, $\overrightarrow{OQ} = \overrightarrow{q}$, then

$$\overrightarrow{p} + \overrightarrow{q} = \overrightarrow{r}.$$

In Fig. 3.6, suppose P is (a, b) and Q is (c, d). The vector $\overrightarrow{OP}$ is

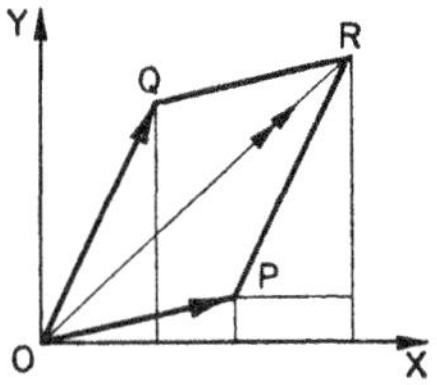

Fig. 3.6

determined by the ordered pair (a, b), so that we can regard (a, b) as a vector. If $\overrightarrow{OQ} = (c, d) = \overrightarrow{PR}$, the relation $\overrightarrow{OP} + \overrightarrow{PR} = \overrightarrow{OR}$ becomes $(a, b) + (c, d) = (a + c, b + d)$.

Ex. 3. Draw other versions of Fig. 3.6, with P and Q in different positions in relation to each other, and in relation to the axes. Verify that, in all cases, $(a, b) + (c, d) = (a + c, b + d)$.

Ex. 4. Give a geometrical interpretation of the argument:

$$(c, d) + (a, b) = (c + a, d + b) = (a + c, b + d),$$

and so

$$(a, b) + (c, d) = (c, d) + (a, b).$$

Is the operation of addition of vectors commutative?

Ex. 5. If $\vec{p}, \vec{q}, \vec{r}$ are vectors, is it true that $\vec{p} + (\vec{q} + \vec{r}) = (\vec{p} + \vec{q}) + \vec{r}$? Is the operation of addition of vectors associative?

Ex. 6. Is there a zero vector, that is, is there a vector, 0, such that $\vec{p} + 0 = 0 + \vec{p} = \vec{p}$?

Ex. 7. Is there an additive inverse to the vector $\vec{p}$, that is, a vector $(-\vec{p})$ such that $\vec{p} + (-\vec{p}) = (-\vec{p}) + \vec{p} = 0$?

Subtraction of vectors

The vector $\overrightarrow{OP'}$, equal in length but opposite in direction to $\overrightarrow{OP}$, is denoted by $-\overrightarrow{OP}$, or $-\vec{p}$. This is illustrated in Fig. 3.7

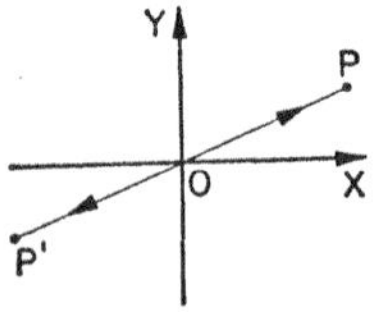

Fig. 3.7

By $\vec{p} - \vec{q}$ we mean $\vec{p} + (-\vec{q})$. In Fig. 3.8, $\overrightarrow{OP} = \vec{p}$, $\overrightarrow{OQ} = \vec{q}$.

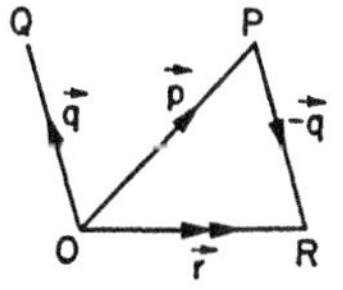

Fig. 3.8

PR is equal and parallel to QO, so that $\overrightarrow{PR} = -\vec{q}$, and $OQPR$ is a parallelogram. Then $\vec{p} - \vec{q} = \vec{p} + (-\vec{q}) = \vec{r}$, where $\vec{r} = \overrightarrow{OR}$. But QP is equal and parallel to OR, and so

$$\vec{p} = \vec{q} + \vec{r}.$$

It follows that $\vec{p} = \vec{q} + \vec{r} \Leftrightarrow \vec{p} - \vec{q} = \vec{r}$.

Ex. 8. If $\vec{p} = (a, b)$ and $\vec{q} = (c, d)$,

(i) express $-\vec{q}$ as an ordered pair, (ii) express $\vec{p} - \vec{q} = \vec{r}$ in terms of ordered pairs.

Scalar multiplier

Suppose $\overrightarrow{OP} = \vec{r}$ and let k be any real number, not a vector. In Fig. 3.9, let $OQ = k \, . \, OP$, so that $\overrightarrow{OQ}$ is a vector with the same

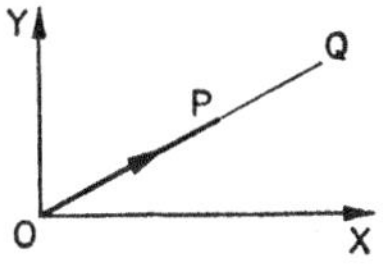

Fig. 3.9

direction as $\overrightarrow{OP}$ and with a length k times the length of $\overrightarrow{OP}$. We write

$$\overrightarrow{OQ} = k \, . \, \overrightarrow{OP} \quad \text{or} \quad \vec{q} = k\vec{p}.$$

If k is negative, the direction of $\overrightarrow{OP}$ is reversed; thus, in Fig. 3.10,

$$\overrightarrow{OQ} = -\tfrac{1}{2}\overrightarrow{OP}.$$

The multiplier k, a real number, is called a *scalar*, and the operation is called *scalar multiplication.*

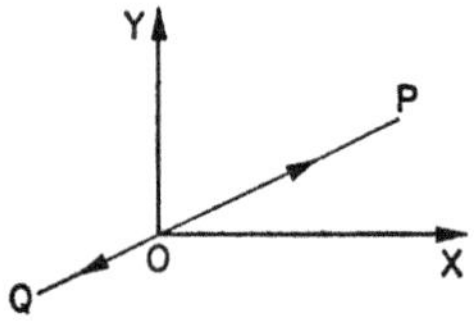

Fig. 3.10

Ex. 9. If a, b are scalars and $\overrightarrow{u}, \overrightarrow{v}$ are vectors, verify that (i) $a \,.\, \overrightarrow{u}$ is a vector (Closure); (ii) $1 \,.\, \overrightarrow{u} = \overrightarrow{u}$ (Unit scalar); (iii) $a(b \,.\, \overrightarrow{u}) = (ab) \,.\, \overrightarrow{u}$ (Associative); (iv) $a \,.\, (\overrightarrow{u} + \overrightarrow{v}) = a \,.\, \overrightarrow{u} + a \,.\, \overrightarrow{v}$, $(a + b) \,.\, \overrightarrow{u} = a \,.\, \overrightarrow{u} + b \,.\, \overrightarrow{u}$ (Distributive).

It is now clear that we have an unusual set of operations. Vectors alone, under the operation of addition, form a closed system following the commutative and associative laws, and have a zero and an additive inverse. Vectors under scalar multiplication also form a closed system following modified laws of association and distribution, with a scalar zero and a scalar unit. We shall now examine some of these properties when vectors are presented as ordered pairs.

Suppose P is (a, b), Q is (c, d) and $\overrightarrow{OQ} = k \,.\, \overrightarrow{OP}$. Then $c = ka$, $d = kb$, and so

$$k \,.\, (a, b) = (ka, kb).$$

Mid-points

Let P_1, P_2 be (x_1, y_1), (x_2, y_2) and suppose $P(x, y)$ is the mid-point of P_1P_2.

Then

$$\overrightarrow{P_1P} = \overrightarrow{PP_2},$$

and so
$$\overrightarrow{OP} - \overrightarrow{OP_1} = \overrightarrow{OP_2} - \overrightarrow{OP}$$

or

$$\vec{OP} = \tfrac{1}{2}.(\vec{OP_1} + \vec{OP_2}),$$

$$= \tfrac{1}{2}.\vec{OP_1} + \tfrac{1}{2}.\vec{OP_2}.$$

Hence

$$(x, y) = \tfrac{1}{2}.(x_1, y_1) + \tfrac{1}{2}.(x_2, y_2)$$

$$= \{\tfrac{1}{2}(x_1 + x_2),\ \tfrac{1}{2}(y_1 + y_2)\}.$$

Therefore

$$x = \tfrac{1}{2}(x_1 + x_2), \qquad y = \tfrac{1}{2}(y_1 + y_2).$$

The coordinates of the mid-point of P_1P_2 are therefore

$$\left(\frac{x_1 + x_2}{2}, \frac{y_1 + y_2}{2}\right).$$

The section formula

Let P divide P_1P_2 in the ratio $a : b$ (Fig. 3.11), so that

$$b.\vec{P_1P} = a.\vec{PP_2}.$$

Then

$$b(\vec{p} - \vec{p_1}) = a(\vec{p_2} - \vec{p}),$$

$$(a + b)\vec{p} = b\vec{p_1} + a\vec{p_2}.$$

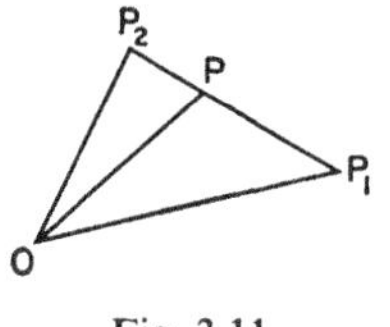

Fig. 3.11

Hence

$$(a + b).(x, y) = b.(x_1, y_1) + a.(x_2, y_2),$$

or

$$\{(a + b)x, (a + b)y\} = \{(bx_1 + ax_2), (by_1 + ay_2)\},$$

giving

$$(a + b)x = bx_1 + ax_2, \quad (a + b)y = by_1 + ay_2.$$

The coordinates of the point dividing P_1P_2 in the ratio $a : b$ are therefore

$$\left(\frac{bx_1 + ax_2}{a + b}, \frac{by_1 + ay_2}{a + b}\right).$$

Ex. 10. If $5\overrightarrow{OP} = 2\overrightarrow{OQ} + 3\overrightarrow{OR}$, prove that $2\overrightarrow{QP} = 3\overrightarrow{PR}$. What does this tell us about the position of P in relation to Q and R?

Ex. 11. If $\overrightarrow{OP} = \vec{p}$, $\overrightarrow{OQ} = \vec{q}$, $\overrightarrow{OR} = \vec{r}$ and $(\lambda + \mu)\vec{p} = \lambda\vec{q} + \mu\vec{r}$, what can we deduce about the position of P in relation to Q and R?

Homothetic figures

A *locus* is a set of points. For example, if P is (x, y), the locus $\{(x, y)|x^2 + y^2 = 1\}$ is the set of points on the circumference of the circle with centre O and radius 1 unit (see Fig. 3.12); the locus $\{(x, y)|x^2 + y^2 \leqslant 1\}$ is the set of points on or within the circle.

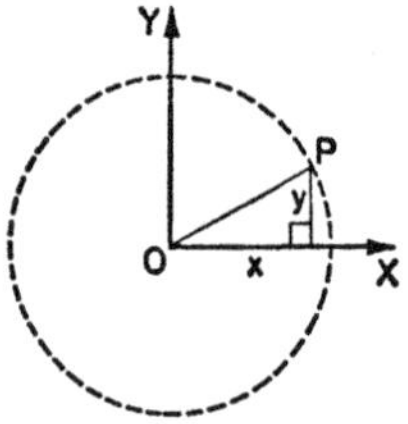

Fig. 3.12

Suppose O is a fixed point and $P(x, y)$ is a member of a locus $\{P\}$. Let k be a fixed scalar number. If Q is a point on OP such that $\overrightarrow{OQ} = k \,.\, \overrightarrow{OP}$, then the locus $\{Q\}$ is said to be *homothetic* with the locus of P. O is called the homothetic centre.

EXAMPLE. *If two polygons $P_1P_2P_3 \ldots$ and $Q_1Q_2Q_3 \ldots$ have their corresponding sides parallel and proportional, they are homothetic.*

Suppose $\overrightarrow{P_1P_2} = k.\overrightarrow{Q_1Q_2}$, and let P_1Q_1, P_2Q_2 meet at O (see Fig. 3.13). Then

$$\overrightarrow{OP_2} - \overrightarrow{OP_1} = k\,.\,(\overrightarrow{OQ_2} - \overrightarrow{OQ_1}).$$

Let $OP_1 = \lambda\,.\,OQ_1$, $OP_2 = \mu\,.\,OQ_2$; then

$$(\mu - k)\,.\,\overrightarrow{OQ_2} = (\lambda - k)\,.\,\overrightarrow{OQ_1}.$$

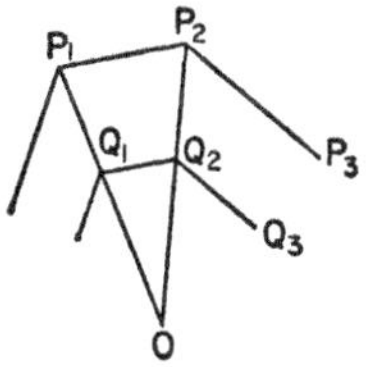

Fig. 3.13

Since OQ_1 and OQ_2 are not parallel, this relation is impossible unless $\mu = k$, $\lambda = k$. Hence $OP_1 = k\,.\,OQ_1$ and $OP_2 = k\,.\,OQ_2$. Similarly P_3Q_3 will meet P_2Q_2 at O where $OP_3 = k\,.\,OQ_3$. Hence $\{P_1P_2P_3\ldots\}$ and $\{Q_1Q_2Q_3\ldots\}$ are homothetic.

EXERCISE 3.1

1. If $\overrightarrow{OP} = (-2, 6)$, $\overrightarrow{OQ} = (8, -4)$, write down the ordered pairs equivalent to:

(i) $\overrightarrow{PQ}$, (ii) $\overrightarrow{QP}$, (iii) $\overrightarrow{OP} + \overrightarrow{OQ}$, (iv) $\overrightarrow{OP} + \overrightarrow{PQ} + \overrightarrow{QO}$, (v) the mid-point of PQ, (vi) the point which divides PQ in the ratio 3:2.

2. Draw sketches to illustrate the loci:
(i) $A = \{(x, y)|-1 < x < 1\}$, (ii) $B = \{(x, y)|0 < y < 2\}$, (iii) $A \cup B$, (iv) $A \cap B$.

3. Write down the coordinates of the mid-points of the lines joining (i) (1, 2) and (4, 8), (ii) (2, 5) and (3, 5). What type of quadrilateral has its vertices at the points given in (i) and (ii)?

4. Describe the position of P in relation to Q and R when $\overrightarrow{OP} = 2\overrightarrow{OQ} - \overrightarrow{OR}$.
Prove that (1, 3), (5, −1), (−3, 7) are collinear.

5. The vertices of a triangle ABC are $(5, -7)$, $(-2, 11)$, $(6, -1)$. Write down the coordinates of A', B', C', the mid-points of BC, CA, AB.

Find the coordinates of the points which divide the medians AA', BB', CC' in the ratio 2:1. What can be said about the medians of the triangle?

6. Show that the triangle formed by the mid-points of the sides of a triangle is homothetic with the given triangle.

What is the radius of the circle through the mid-points of the sides of a triangle whose circumradius is R?

7. A *centroid* of the n points $P_1, P_2, \ldots P_n$ is defined to be the point G where $n \,.\, \vec{OG} = \vec{OP_1} + \vec{OP_2} + \ldots + \vec{OP_n}$. Find the centroid of the three points A, B, C of no. 5. Compare this with the answers to no. 5.

8. Find the centroid of the four points O, A, B, C where O is $(0, 0)$, A is $(-4, 5)$, B is $(2, 1)$, C is $(2, -1)$. Is this point the same as (i) the mid-point of either diagonal of $OABC$, or (ii) the mid-point of either of the lines joining the mid-points of OA, BC or AB, CO?

9. Show by shading in a figure the loci:
(i) $A = \{(x, y) | y \leqslant x\}$, (ii) $B = \{(x, y) | x + y \leqslant 4\}$, (iii) $A \cup B$, (iv) $A \cap B$. Describe the locus $A \cap B \cap C$, where $C = \{(x, y) | y \geqslant 0\}$.

Three dimensions

The position of a point in space can be referred to three mutually perpendicular planes which meet in three mutually perpendicular axes OX, OY, OZ.

If PL, PM, PN are the distances of P from the planes YOZ, ZOX, XOY (see Fig. 3.14), there is a one–one correspondence between the set of points $\{P\}$ in space and the ordered triple (x, y, z), where $PL = x$, $PM = y$, $PN = z$. If $u = (x, y)$ is an

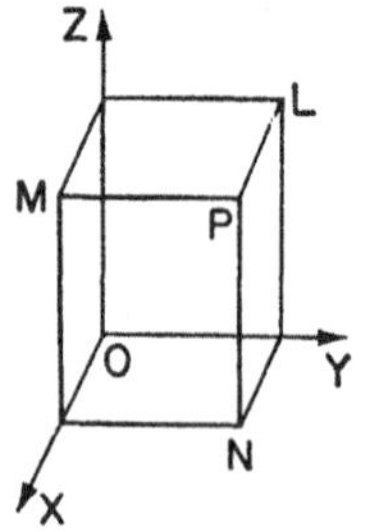

Fig. 3.14

ordered pair, then the ordered triple (x, y, z) is equivalent to the ordered pair (u, z).

The displacement $\overrightarrow{OP}$ and the ordered triple (x, y, z) are different aspects of the same vector $\vec{p}$. If $\vec{p_1} = \overrightarrow{OP_1} = (x_1, y_1, z_1)$ and $\vec{p_2} = \overrightarrow{OP_2} = (x_2, y_2, z_2)$, then

$$\vec{p_2} = k \,.\, \vec{p_1} \Leftrightarrow \overrightarrow{OP_2} = k \,.\, \overrightarrow{OP_1}$$
$$\Leftrightarrow (x_2, y_2, z_2) = k \,.\, (x_1, y_1, z_1).$$

Exactly as in two dimensions we have

$$\overrightarrow{OP} + \overrightarrow{PQ} = \overrightarrow{OQ},$$

$$(x_1, y_1, z_1) = (x_2, y_2, z_2) \Leftrightarrow x_1 = x_2, y_1 = y_2, z_1 = z_2,$$
$$k \,.\, (x_1, y_1, z_1) = (kx_1, ky_1, kz_1).$$

Ex. 12. If P is the mid-point of the line joining $P_1(x_1, y_1, z_1)$ and $P_2(x_2, y_2, z_2)$, prove that the coordinates of P are

$$\left(\frac{x_1 + x_2}{2}, \frac{y_1 + y_2}{2}, \frac{z_1 + z_2}{2}\right).$$

Ex. 13. If P divides P_1P_2 in the ratio $a : b$, prove that the coordinates of P are

$$\left(\frac{bx_1 + ax_2}{a + b}, \frac{by_1 + ay_2}{a + b}, \frac{bz_1 + az_2}{a + b}\right).$$

EXAMPLE. *A, B, C, D are four collinear points, and O is a point not on AD. A second line cuts OA, OB, OC, OD in P, Q, R S. Prove that*

$$\frac{AB}{BC}\Big/\frac{AD}{DC} = \frac{PQ}{QR}\Big/\frac{PS}{SR}.$$

Let $\overrightarrow{OA} = \vec{a}$, $\overrightarrow{OP} = \vec{p}$, and so on, and suppose

$$\vec{p} = \lambda \,.\, \vec{a}, \quad \vec{r} = \mu \,.\, \vec{c} \quad \text{(see Fig. 3.15).}$$

If $AB/BC = k_1/1$, $PQ/QR = k_2/1$,

$$\vec{b} = \frac{\vec{a} + k_1 . \vec{c}}{1 + k_1}, \quad \vec{q} = \frac{\vec{p} + k_2 . \vec{r}}{1 + k_2}.$$

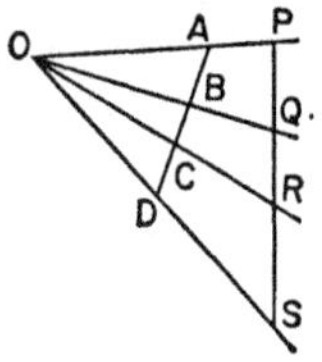

Fig. 3.15

But $\vec{q} = \alpha . \vec{b}$, where α is a scalar.

Hence

$$\frac{\lambda . \vec{a} + k_2\mu . \vec{c}}{1 + k_2} = \frac{\alpha . \vec{a} + \alpha k_1 . \vec{c}}{1 + k_1}.$$

It follows that

$$\frac{\lambda}{1 + k_2} = \frac{\alpha}{1 + k_1}, \quad \frac{k_2\mu}{1 + k_2} = \frac{\alpha k_1}{1 + k_1},$$

and so

$$\lambda/k_2\mu = 1/k_1,$$

or

$$k_1/k_2 = \mu/\lambda.$$

Similarly, if $AD/DC = k_3/1$ and $PS/SR = k_4/1$, then

$$k_3/k_4 = \mu/\lambda.$$

Hence

$$\frac{AB}{BC}\bigg/\frac{AD}{DC} = \frac{k_1}{k_3} = \frac{k_2}{k_4} = \frac{PQ}{QR}\bigg/\frac{PS}{SR}.$$

Note. $\frac{AB}{BC}\Big/\frac{AD}{DC}$ is called the *cross-ratio* of the set $\{A, B, C, D\}$, and we have shown that the set $\{P, Q, R, S\}$ has the same cross-ratio. This property is the basis of *projective geometry*.

EXERCISE 3.2

1. Let $\overrightarrow{OP} = (2, -3, 5)$, $\overrightarrow{OQ} = (-2, 7, 3)$. Write down ordered triples to represent:

(i) $\overrightarrow{PQ}$, (ii) $\overrightarrow{QP}$, (iii) $\overrightarrow{OP} + \overrightarrow{OQ}$, (iv) $\overrightarrow{OP} + \overrightarrow{PQ} + \overrightarrow{QO}$, (v) the mid-point of PQ, (vi) the point which divides PQ in the ratio 3:2.

2. Describe the loci:
(i) $A = \{(x, y, z) | 0 < x < 1\}$, (ii) $B = \{(x, y, z) | 0 < y < \frac{1}{2}\}$,
(iii) $C = \{(x, y, z) | 0 < z < \frac{1}{3}\}$, (iv) $A \cap B \cap C$.

3. The vertices A, B, C, D of a tetrahedron are $(-1, 2, 3)$, $(3, -1, 4)$, $(5, 3, -2)$, $(3, 4, 1)$. P, Q, R, S, U, V are the mid-points of AB, BC, CD, AD, AC, BD; write down the coordinates of these six points.

L, M, N are the mid-points of PR, QS, UV. Write down the coordinates of these three points. What can be said about the lines joining the mid-points of the opposite sides of $ABCD$? Write down also the coordinates of the centroid of A, B, C, D. (See Exercise 3.1, no. 7.)

4. Describe the position of P in relation to Q and R when

$$2\overrightarrow{OP} = 3\overrightarrow{OQ} - \overrightarrow{OR}.$$

Prove that the points $(-1, -2, 3)$, $(1, 4, 7)$ and $(-2, -5, 1)$ are collinear.

5. In no. 3 above, the medians of $\triangle BCD$ meet at G_1, the centroid of the triangle. Write down the coordinates of G_1, and of the point which divides AG_1 in the ratio 3:1.

Repeat this process for G_2, the centroid of $\triangle ACD$, and BG_2. What can be deduced about the lines joining the vertices of $ABCD$ to the centroids of the opposite faces?

6. P, Q, R, S are the centroids of the faces BCD, ACD, ABD, ABC of the tetrahedron $ABCD$. Show that $\overrightarrow{BA} = 3\overrightarrow{PQ}$, and deduce that $ABCD$ and $PQRS$ are homothetic.

What information does this give about AP, BQ, CR, DS?

Congruences

The time shown by a 12-hour clock never exceeds 12 hours, reverting repeatedly to zero and turning on again towards 12. A similar effect is observed if we divide the integers in turn by a fixed number and write down the remainder. For example, if the integers starting with zero are divided by 5, the successive remainders are 0, 1, 2, 3, 4, 0, 1, 2, 3, 4, 0, ...

We say that the set of numbers {0, 1, 2, 3, 4} is the set of remainders *modulo* 5. In general, the set of remainders modulo n is $\{0, 1, 2, \ldots (n-1)\}$; the divisor n is called the *modulus*.

Two numbers are said to be *congruent* modulo n if they have the same remainder when divided by n; thus -4, 11, 31 are all congruent modulo 5. We write

$$-4 \equiv 11 \pmod 5 \quad \text{or} \quad 31 \equiv 1 \pmod 5.$$

Alternatively, if the integers a and b are congruent modulo m, we can write $a - b \equiv 0 \pmod m$.

Definition: An integer b is divisible by another integer a if there exists another integer d such that $b = ad$, and we then write $a|b$ (read a divides b).

Ex. 14. $a \equiv b \pmod m$ if and only if $m|(a-b)$.

Suppose $a - b = cm$, where c is an integer. Divide b by m, giving a quotient q and a remainder r, less than m; that is

$$b = mq + r, \qquad 0 \leqslant r < m.$$

But $a = b + cm$, and so $a = (c + q)\, m + r$. Hence, if $m|(a - b)$, then a and b have the same remainder when divided by m.

Conversely, if a and b have the same remainder when divided by m,

$$a = q_1 m + r \text{ and } b = q_2 m + r \quad \Rightarrow \quad a - b = (q_1 - q_2)\, m,$$

so that, if a and b have the same remainder when divided by m, then $m|(a - b)$.

The two statements, (i) $m|(a - b)$, and (ii) a and b have the same remainder when divided by m, are therefore equivalent. In either case we write

$$a \equiv b \pmod m \quad \text{or} \quad a - b \equiv 0 \pmod m.$$

Consider the set of *residues*, or remainders, modulo 5: $N = \{0, 1, 2, 3, 4\}$. We define the sum and the product of two residues (mod 5) to be the residue to which the sum or product is congruent. For example,

$$2 + 3 = 5 \equiv 0 \pmod 5, \qquad 2 \times 3 = 6 \equiv 1 \pmod 5.$$

Division is defined as the inverse of multiplication; for example, since $2 \times 3 \equiv 1 \pmod 5$, we have

$$1/3 \equiv 2 \pmod 5, \qquad 1/2 \equiv 3 \pmod 5.$$

Division by 0 is, of course, excluded, but so is division by 5, since $5 \equiv 0 \pmod 5$.

Ex. 15. Construct addition and multiplication tables for the set N of residues modulo 5.

Ex. 16. Use the tables modulo 5 to find the value of:
(i) $1 - 3$, (ii) $0 - 2$, (iii) $3 - 4$, (iv) $2/3$, (v) $3/2$, (vi) $3/4$,
(vii) $2(3 + 4)$, (viii) $2.3 + 2.4$, (ix) $(3 - 4)/(2 - 4)$.

The addition and multiplication tables modulo 5 can be used to verify the truth of the laws of Chapter 1, p. 12, modified below for congruences. These laws will be found to be true for any prime modulus. (A prime number n has no integral factors other than 1 and n.) Law no. 5 fails for moduli which are not prime; for example, there is no number a such that $2a \equiv 1 \pmod 6$.

1. **Closure**

If $a, b \in \{0, 1, 2, 3, \ldots (n - 1)\}$, then $a + b$ and ab, reduced modulo n, are elements of the same set.

2. **Commutative**

$$a + b \equiv b + a, \qquad ab \equiv ba.$$

3. **Associative**

$$a + (b + c) \equiv (a + b) + c, \qquad a(bc) \equiv (ab)c.$$

4. **Identity**

Numbers 0 and 1 exist such that

$$a + 0 \equiv 0 + a \equiv a, \qquad a \times 1 \equiv 1 \times a \equiv a.$$

5. Inverse

Corresponding to a there exist numbers denoted by $(-a)$ and a^{-1} such that:

$$a + (-a) \equiv (-a) + a \equiv 0,$$

$$a \times (a^{-1}) \equiv (a^{-1}) \times a \equiv 1 \pmod{n},$$

provided that n is prime and $a \not\equiv 0 \pmod{n}$.

6. Distributive

$$a(b + c) \equiv ab + ac, \qquad (a + b)\,c \equiv ac + bc.$$

The rule of cancellation for addition presents no difficulty, since, for any integer x,

$$a + x \equiv b + x \Leftrightarrow (a + x) - (b + x) \equiv 0 \Leftrightarrow a - b \equiv 0$$

and so

$$a + x \equiv b + x \quad \Leftrightarrow \quad a \equiv b.$$

There is, however, a difficulty in the rule of cancellation for products when the modulus is not a prime number. For example;

$$2a \equiv 2b \pmod{6} \Leftrightarrow 2(a - b) = 6k,$$

from which we can deduce only that $a \equiv b \pmod{3}$, and from $6a \equiv 6b \pmod{8}$ we can deduce only that $3a \equiv 3b \pmod{4}$.

EXAMPLE. *Solve the equation* $4x + 2 \equiv 3 \pmod{5}$.

$$4x + 2 \equiv 3 \pmod{5} \Leftrightarrow 4x - 1 \equiv 0 \pmod{5}.$$

Hence

$$4x = 5a + 1, \text{ where } a \text{ is an integer.}$$

Let

$$a = 4b + c, \text{ where } b \text{ and } c \text{ are integers.}$$

Then

$$4x = 20b + 5c + 1,$$

$$x = 5b + c + \tfrac{1}{4}(c + 1).$$

Taking $c = 3$,

$$x \equiv 4 \pmod{5}.$$

In some cases congruence equations have more than one solution.

EXAMPLE. *Solve the equation* $6x \equiv 4 \pmod{14}$.

$6x \equiv 4 \pmod{14} \Leftrightarrow 6x = 14a + 4$, where a is an integer.

Let $a = 6b + c$, where b and c are integers.

Then

$$6x = 14\,(6b + c) + 4,$$

$$x = 14b + 2c + \tfrac{1}{6}\,(2c + 4).$$

Possible values of c for which $2c + \frac{1}{3}\,(c + 2) < 14$ are $c = 1, 4$, giving $x \equiv 3, 10 \pmod{14}$.

EXAMPLE. *Prove that* $3 \,.\, 5^5 + 2^7$ *is divisible by* 17.

$$\begin{aligned} 3 \,.\, 5^5 + 2^7 &= 15 \,.\, 25^2 + 4 \,.\, 32 \\ &= (17 - 2)\,(17 + 8)^2 + 128 \\ &\equiv -2 \,.\, 64 + 128 \pmod{17}. \end{aligned}$$

Hence

$$3 \,.\, 5^5 + 2^7 \equiv 0 \pmod{17}.$$

EXERCISE 3.3

1. Make addition and multiplication tables for the set of residues modulo 7. Hence evaluate (mod 7):

(i) $3 - 6$, (ii) $1 - 5$, (iii) $3/6$, (iv) $1/2$, (v) $(2 - 4)/(1 - 6)$, (vi) $3\frac{4}{5}$.

Solve the congruences in nos. 2–10:

2. $3x \equiv 1 \pmod{4}$.

3. $3x \equiv 5 \pmod{7}$.

4. $4x + 1 \equiv 7 \pmod{10}$.

5. $2x - 1 \equiv 8 \pmod{9}$.

6. $5x + 7 \equiv 2 \pmod{10}$.

7. $3x - 5 \equiv 7 \pmod{15}$.

8. $4x + 5 \equiv 3 \pmod{10}$.

9. $8x - 1 \equiv 4 \pmod{23}$.

10. $2x - 1 \equiv 2 \pmod{4}$.

11. By considering the results of nos. 2–10, state the conditions for the congruence $ax + b \equiv c \pmod{m}$ to have (i) a unique solution, (ii) no solution.

12. If n is an integer, find possible values of
(i) $n^2 \pmod{5}$, (ii) $n^2 \pmod{10}$, (iii) $n^3 \pmod{10}$.

13. Prove that $3^8 - 1$ is divisible by 80.

14. Prove that $3 . 4^8 + 10^6 \equiv 4 \pmod{9}$.

15. Show that the product of four consecutive integers is divisible by 24, and deduce that

$$n(n + 23)(n + 25)(n + 74) \equiv 0 \pmod{24}.$$

16. Solve, where possible, the congruences modulo 10:
(i) $n^2 \equiv 1$, (ii) $n^2 \equiv 2$, (iii) $n^2 \equiv 4$, (iv) $n^2 \equiv 6$, (v) $n^2 \equiv 9$.

17. If $a \equiv b \pmod{m}$, show that for all integers n:
(i) $a + n \equiv b + n \pmod{m}$, (ii) $an \equiv bn \pmod{m}$.
Under what conditions is the converse of (ii) true?

CHAPTER 4

FUNCTIONS

Relations

A *relation* is a means of establishing a bond between the elements of two sets, or between elements of the same set. It is part of a sentence such as

"... is the son of ...",

"... is less than ...",

"... is twice as large as ...".

Such a bond between two elements establishes an ordered pair. If R represents the relation which, with x as the first element, selects the second element y, we can write yRx.

Figure 4.1 is a graphical representation of a relation leading from elements of a set A to elements of a set B.

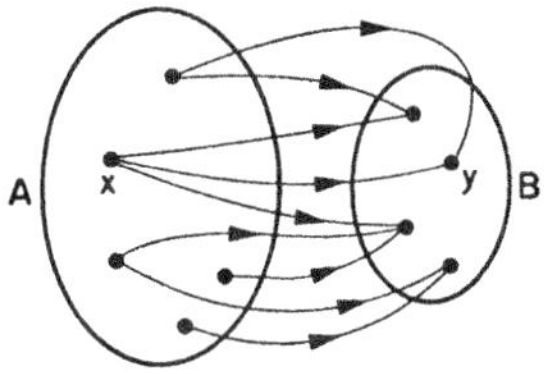

Fig. 4.1

Suppose A represents a set of boys and B the set of subjects that these boys can study at school. R is the relation

"... is studying ...".

An arrow from a point in A, representing a boy, to a point in B, representing a subject, indicates that the boy is studying the subject. The figure can be called the graph of the relation.

Alternatively the same relation can be shown in the form of a table, as in Fig. 4.2; a mark in a square of the table corresponds to an arrow in the graph.

Name (x)	Subject (y)			
	a	b	c	d
Brown	X	X		
Hughes	X	X	X	
Smith			X	X
Taylor			X	
Williams				X

Fig. 4.2

When the elements of an ordered pair are numbers the relation between them can be represented by a point referred to axes of coordinates OX, OY. Suppose the relation is "y is less than x", between the elements of the sets $X = \{1, 3, 5, 7\}$ and $Y = \{1, 2, 3, 4\}$. The relation yRx gives the ordered pairs

$$(3, 1), \quad (3, 2), \quad (5, 1), \quad (5, 2), \quad (5, 3), \quad (5, 4),$$
$$(7, 1), \quad (7, 2), \quad (7, 3), \quad (7, 4).$$

The corresponding points are shown in Fig. 4.3.

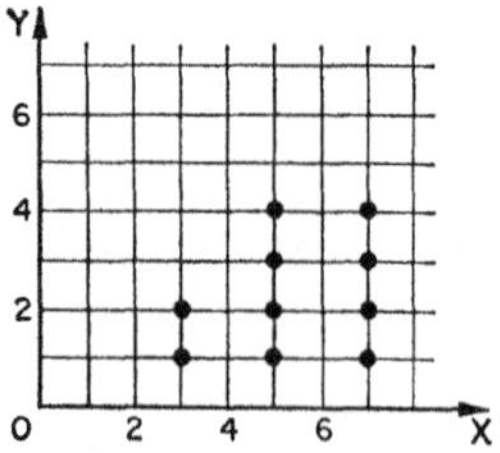

Fig. 4.3

The cartesian product $X \times Y$ contains sixteen ordered pairs, from which the relation R selects a subset of ten pairs. Conversely, every proper subset of $X \times Y$ constitutes a relation between the elements of X and Y. Failing a concise statement or a formula, the plotting of points as in Fig. 4.3 is an effective way of showing the relation.

Mapping

In Fig. 4.4, **A** represents a set of boys and **B** represents a set of positive integers. The graph shows the relation R:

"The present age of . . . in years is . . .".

Since a boy cannot have two different ages at the same time, each point of **A** is the origin of one and only one arrowed line. In such a case we say that the relation R *maps the set* **A** *into the set* **B**.

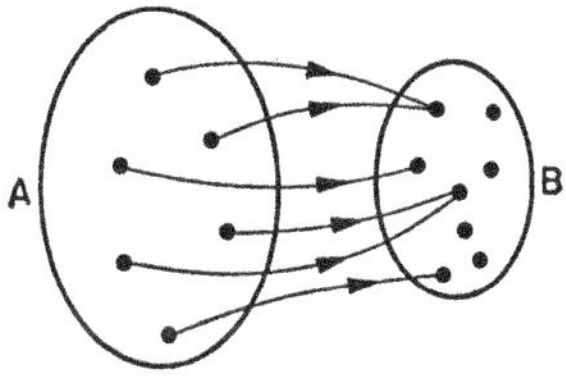

Fig. 4.4

Suppose we have a relation R which maps a set **A** into a set **B**, and that an element x of **A** gives rise to an element y of **B**. Then y is called the *image* of x by R, and we write

$$y = R(x).$$

The set of images of all the elements of **A** by R is called the image of **A** and is denoted by $R(\mathbf{A})$. If $R(\mathbf{A}) = \mathbf{B}$, so that every element of **B** is an image of some element of **A** by R, we say that R maps **A** *onto* **B**.

The concept of a relation is very broad. Any proper subset of the cartesian product $A \times B$ constitutes a relation from A to B, and the subset can be taken as defining the relation R. Thus R, considered as a set, should be distinguished from its statement in words, which is written yRx. In symbols:

$$R = \{(x, y) | x \in A, y \in B, yRx\}.$$

It is not implied that every element of **A** has a corresponding element in **B**. To take a numerical example, $y^2 = x$ requires that x must be positive or zero. Those elements of **A** which have an image in **B** under R constitute a subset of **A** which is called the *domain* of R. Similarly there may be elements of **B** which are not images of any x in **A**; those which are images of an x constitute a subset of **B** which is called the *range* of R.

We have just seen that a mapping of **A** into **B** requires that to each $x \in$ **A** there is a unique image $y \in$ **B**, but, in general, there may be several values of x which have the same image in **B**.

Ex. 1. If $X = \{1, 2, 3\}$, $Y = \{1, 2, 3, 4, 5, 6\}$, how many elements are there in $X \times Y$? Which of these elements belong to the set

(i) $\{(x, y) | y = 2x\}$, (ii) $\{(x, y) | y = x^2\}$?

Ex. 2. If $X = Y = \{0, 1, 2, 3, 4, 5\}$, write down the relation R whose graph is given in Fig. 4.5. What are the domain and range of R?

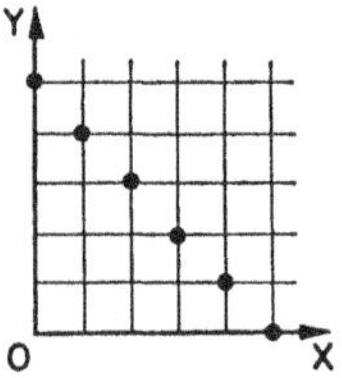

Fig. 4.5

Ex. 3. In Fig. 4.6, suppose A is (5, 0) and B is (0, 10). Write down a description of the sets X and Y, and of the relation R whose graph is the shaded area. What are the domain and range of R?

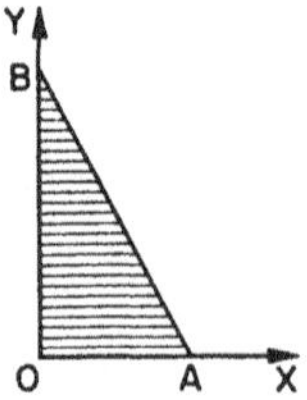

Fig. 4.6

Functions

In general, mappings are not concerned with numbers only. If **A** is a set of boys and **B** is a set of men, the relation x *is the son of* y leads to a mapping from **A** into **B**. When numbers form the elements of the sets the word *function* is commonly used to describe the mapping, although its use is not confined to numerical mappings.

Suppose $x \in X$, $y \in Y$. Then y is a function of x, which we write $y = f(x)$, if, when x is given, y is uniquely determined; x is called the *independent variable* and y the *dependent variable*. The value of y when $x = a$ is denoted by $f(a)$.

For example, $y = x^2 + 2$ defines y as a function of x, and $f(2) = 2^2 + 2 = 6$.

In the language of sets, a function f from the set X to the set Y is the set of ordered pairs (x, y) defined by the relation yfx, and is such that, if $y \in Y$ exists for a given $x \in X$, then y is uniquely determined.

The relation "y is a function of x" is symbolized in many ways by different writers. In additon to $y = f(x)$ and yfx the reader may meet, amongst others,

$$f : X \to Y, \qquad \underset{\curvearrowright}{xfy}, \qquad y = xf.$$

Another symbol which is commonly used is the *quantifier* $\exists$, which stands for "there exists . . . such that", or "we can find a value of . . . such that"; for example, $\exists x > 3$ represents "we can find a value of x greater than 3". In the formal language of sets the definitions of function, domain and range are as follows:

(i) f is the set $\{(x, y) | x \in X, y \in Y, y = f(x)\}$;

(ii) the domain of f is $\{x | \exists y : y = f(x)\}$;

(iii) the range of f is $\{y | \exists x : y = f(x)\}$.

Strictly then, f is a subset of the cartesian product $X \times Y$ and should be distinguished from the relation yfx, but the distinction is rarely important. The domain of f is a subset of X, and the range of f is a subset of Y.

Ex. 4. If $f(x) = (x - 1)/(x + 2)$, find the value of $f(1), f(2), f(-3), f(\frac{2}{3})$ $f(x + 1)$.

Ex. 5. If $f(x) = (x - 1)^2 + 2$, find the value of $f(-2), f(1), f(3), f(\frac{1}{2})$, $f(2x + 1)$.

What is the range of $f(x)$?

Ex. 6. If $f(x) = \sqrt{\{(x - 1)(2 - x)\}}$, find the vealu of $f(1), f(\frac{3}{2}), f(x + 1)$, $f(x + 2)$.

What is the domain of $f(x)$?

Inverse functions

Figure 4.7 represents a mapping from the set $\{X\}$ into the set $\{Y\}$. An element $y \in Y$ can be the image of more than one element $x \in X$. The relation in which x is given and y is consequently determined is a function and we can write $y = f(x)$. Thus, for a given $x_1 \in X$, if $y_1 = f(x_1)$ and $y_2 = f(x_1)$, then $y_1 = y_2$ but $y_1 = f(x_1)$ and $y_1 = f(x_2)$ does not necessarily imply $x_1 = x_2$; in other words, x is not a function of y.

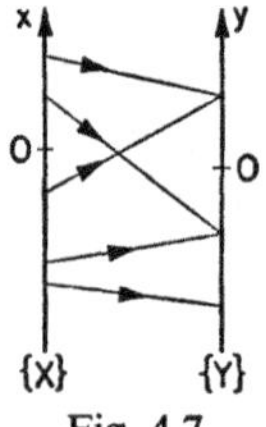

Fig. 4.7

Figure 4.8 represents another mapping, in this case from Y into X. In this case

$$(x_1, y_1) = (x_2, y_1) \Rightarrow x_1 = x_2,$$

and x is a function of y, but y is not a function of x.

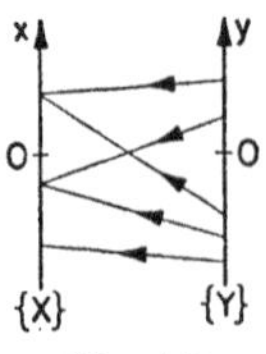

Fig. 4.8

Figure 4.9 shows an example of a one–one relation. If the relation from x to y is $y = f(x)$, the relation from y to x is written $x = f^{-1}(y)$ and is called the *inverse function* of f.

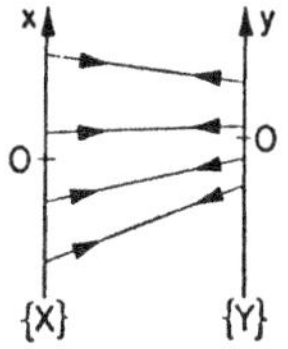

Fig. 4.9

When f^{-1} exists, its domain is the range of f, and its range is the domain of f.

A simple example of a function and its inverse is

$$f = \{(x, y)|y = 2x + 1\},$$
$$f^{-1} = \{(y, x)|x = \tfrac{1}{2}(y - 1)\}.$$

It should be noted that f^{-1} exists only when the mapping of X into Y is one–one.

It is customary to retain x as the independent and y as the dependent variable even when inverse functions are described; thus, if the given function is $y = 2x + 1$, the inverse function would normally be written $y = \tfrac{1}{2}(x - 1)$.

Ex. 7. Write down the inverse functions of the following: (i) $y = 3 - 2x$, (ii) $y = (1/x)$, (iii) $y = (1 - x)/(1 + x)$.

Locus and graph

We have seen that a locus is a set of points (x, y) which represents the set of ordered pairs defined by a relation. When the locus is plotted on graph paper for a numerical relation, the result is the graph of the relation.

Coordinate geometry is the study of loci by the application of algebra to the corresponding relations. Consider, for example, the locus

$$\{(x, y)|x^2 + y^2 = 1\}.$$

Since $+\sqrt{(x^2 + y^2)}$ is the distance of the point (x, y) from the origin, this locus is the set of points $P(x, y)$ such that $OP = 1$. The graph of the locus, or of the relation $x^2 + y^2 = 1$, is the circle with centre O and radius 1.

If an equation $y = f(x)$ gives rise to a certain graph, we say that $y = f(x)$ is the equation of the graph; thus $y = \sqrt{(1 - x^2)}$ is the equation of the semi-circle with centre O and radius 1 that lies in the first and second quadrants. Conversely, if a graph contains all the points (x, y) which satisfy a certain equation, and no other points, we say it is the graph of the equation. For example, the circle with centre O and radius 1 is the graph of the equation $x^2 + y^2 = 1$.

EXAMPLE. *Construct the graph of the locus* $\{(x, y) | y = x\}$ *when* x *and* y *are real numbers.*

Figure 4.10 shows the points of the set $\{(x, y) | y = x\}$ for $x = -3, -2, -1, 0, 1, 2, 3$. These points lie on the line bisecting the angle between Ox and Oy.

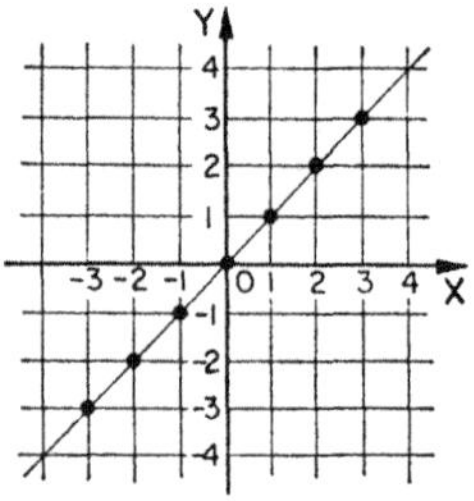

Fig. 4.10

When x, y are real numbers the graph of $\{(x, y) | y = x\}$ is the straight line drawn through the plotted points. We call this line the graph of the equation $x - y = 0$.

Conversely, every point (x, y) on the line bisecting the angle between the axes, and passing through the first and third quadrants, satisfies the equation $x - y = 0$. We say that $x - y = 0$ is the equation of the line.

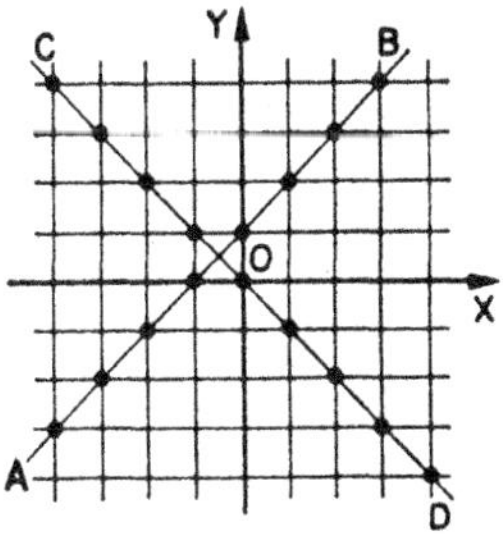

Fig. 4.11

Ex. 8. In Fig. 4.11 the side of each square represents one unit. What are the equations of the lines AB and CD?

Ex. 9. What is the equation of the line passing through the points $(-2, 4)$, $(1, 1)$, $(2, 0)$?

EXAMPLE. *Construct the graph of the locus of* (x, y) *when* $-1 < x < 1$ *and* $0 < y < 1$.

The relation $-1 < x < 1$ specifies the set of points lying between the boundary lines $x = -1$ and $x = +1$. The graph is shown in part in Fig. 4.12 as a horizontally shaded area.

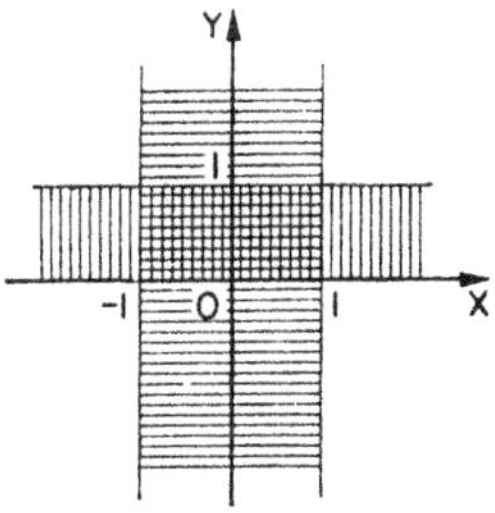

Fig. 4.12

The relation $0 < y < 1$ specifies the set of points lying between the boundary lines $y = 0$ and $y = +1$. The graph is shown in part in Fig. 4.12 as a vertically shaded area.

The locus $\{(x, y)|-1 < x < 1 \text{ and } 0 < y < 1\}$ is the *intersection* of the two sets, and its graph is the doubly shaded area.

Compare with this the graph of

$$\{(x, y) | -1 < x < 1 \text{ or } 0 < y < 1\},$$

which is the *union* of the two sets, and consists of both the shaded areas.

Ex. 10. Construct the graph of the locus of (x, y) when (i) $0 < x < 2$ and $-1 < y < 1$; (ii) $x^2 + y^2 < 1$.

Ex. 11. What is the locus whose graph is shown as the shaded area OAB in Fig. 4.13?

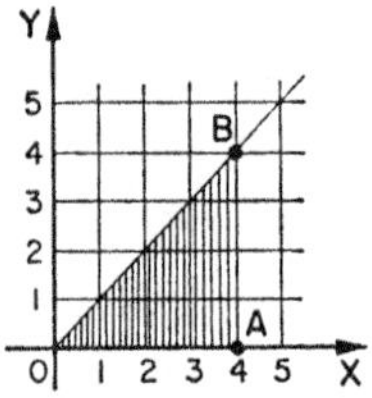

Fig. 4.13

EXAMPLE. *Find the equation of the graph, shown in Fig.* 4.14, *consisting of the lines OA and OB.*

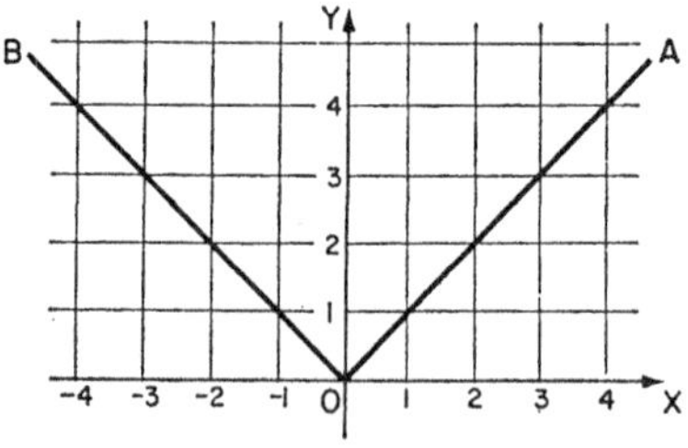

Fig. 4.14

When x is positive the graph is the line OA, whose equation is $y = x$. When x is negative the graph is the line OB, whose equation is $y = -x$. Combining these two we can write either $y = |x|$ or $y = \sqrt{(x^2)}$.

The symbol $|x|$, which is called *modulus* x, indicates that the actual value of x is to be taken with a positive sign; in other words, when $0 < x$, $|x| = +x$, and when $x < 0$, $|x| = -x$. Since $\sqrt{}$ indicates the positive square root of a number, $|x| = \sqrt{(x^2)}$.

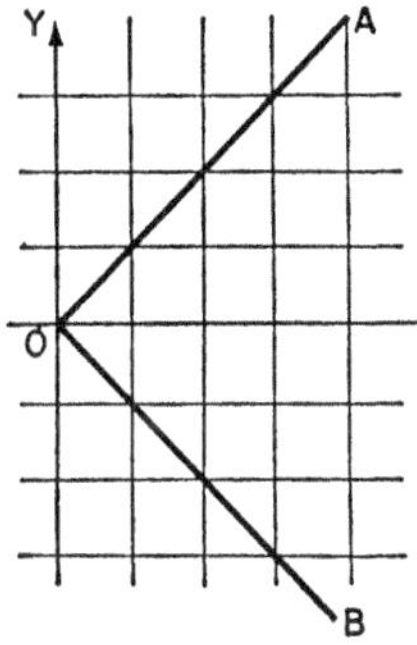

Fig. 4.15

Ex. 12. Find the equation of the graph consisting of the lines *OA* and *OB* in Fig. 4.15.

Ex. 13. Construct the graphs of (i) $y = \sqrt{(x^2)} + 1$; (ii) $y = |x - 1|$.

Ex. 14. (i) In Fig. 4.14, is y a function of x? If so, does the inverse function exist?

(ii) In Fig. 4.15, is either x or y a function of the other?

EXAMPLE. (*i*) *Plot the points of the locus* $\{(x, y)|y = x^2\}$ *for* $x = -3, -2, -1, 0, 1, 2, 3$. (*ii*) *Draw an accurate graph of the function* $y = x^2$ *for values of* x *between* -3 *and* $+3$. *Scales:* 2 *cm to* 1 *unit on x-axis,* 2 *cm to* 2 *units on y-axis.*

(i) The *table of values* of $y = x^2$ is as follows:

x	-3	-2	-1	0	1	2	3
y	9	4	1	0	1	4	9

A sketch of the locus is shown in Fig. 4.16, with equal scales on both axes. The curve is called a *parabola*, and it is symmetrical about the y-axis.

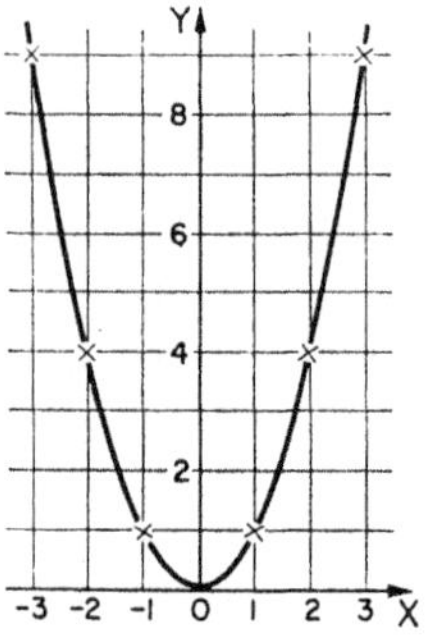

Fig. 4.16

Ex. 15. Sketch the loci: (i) $x > 0, 4 > y, y > x^2$; (ii) $-1 < x < 1$, $-1 < y$, $y < x^2$.

(ii) In order to draw an accurate graph we need extra plotted points. Because of the symmetry, we need to calculate y only for positive values of x. The extra values are shown in the accompanying table.

x	0·5	1·5	2·5
y	0·25	2·25	6·25

The graph of $y = x^2$ is shown in Fig. 4.17. It was originally drawn to a scale of 1 in to 1 unit on OX and 1 in to 2 units on OY, but the scale has had to be reduced in order to fit into this page.

Ex. 16. Use the graph in Fig. 4.17 to read off the squares of $-2·4$, $-1·8$, 0·7, 1·2, 2·3.

Ex. 17. Use the graph in Fig. 4.17 to read off the square roots of 1·3, 2·5, 3·8, 5·9, 7·5.

Ex. 18. Sketch the inverse function of $x > 0, y = x^2$. Has $y = x^2$ an inverse function? Sketch a graph of the equation $y^2 = x$.

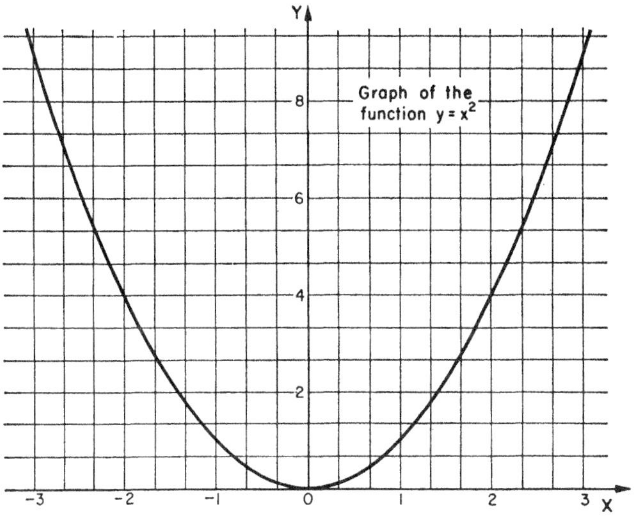

Fig. 4.17

EXERCISE 4.1

1. Draw freehand graphs to illustrate the following sets:
(i) $A = \{(x, y)|x^2 + y^2 \leqslant 1\}$, (ii) $B = \{(x, y)|x \geqslant 0\}$, (iii) $C = \{(x, y)|y \geqslant x\}$, (iv) $A \cup B$, (v) $A \cap C$, (vi) $A \cap B \cap C$.

2. A boy takes an examination in mathematics in which the maximum mark is 200. He is given a grade in accordance with the following scheme, where N is the number of marks he scores:

A	$N \geqslant 185$,	B	$175 \leqslant N < 185$,
C	$158 \leqslant N < 175$,	D	$148 \leqslant N < 158$,
E	$132 \leqslant N < 148$,	F	$119 \leqslant N < 132$,
G	$110 \leqslant N < 119$,	H	$100 \leqslant N < 110$,
I	$70 \leqslant N < 100$,	J	$N < 70$.

Draw a graph to show these gradings, measuring the marks horizontally and the gradings equally spaced vertically.

Is the grade a function of the mark scored? Is the mark scored a function of the grade?

3. A graph has an equation of the form $y = ax^2 + bx + c$. Find a, b, c if the graph passes through the points $(-4, 13)$, $(0, -3)$, $(3, 6)$.

4. Sketch the graphs of (i) $y = x + |x|$, (ii) $y = x - |x|$, (iii) $y = x . |x|$, (iv) $y = x/|x|$.

In each case give the domain and range of the function.

5. Sketch the function $y = x/(1 - x)$ and the inverse function.

6. Sketch the function $y = x^3$ and the inverse function.

7. Suppose f, g are two functions with inverses denoted by f^{-1}, g^{-1}. Define the "product" fg of the two functions to be $f\{g(x)\}$, and show that $(fg)^{-1} = g^{-1}f^{-1}$.

8. Give an example to illustrate each of the following statements, with the notation of no. 7:

(i) $fg = gf$, (ii) $fg \neq gf$.

CHAPTER 5

EXPONENTIAL AND LOGARITHMIC FUNCTION

Notation

As the nature of the numbers in this chapter is important, and since it is essential to know from time to time the type of number being considered, the following notation will be employed:

$\mathbf{J}^+$ is the set of positive integers (or natural numbers),

$\mathbf{J}$ is the set of integers, positive, zero and negative,

$\mathbf{R}$ is the set of rational numbers,

$\mathbf{R}^*$ is the set of real numbers.

The properties of the various types of number were examined in Chapter 1.

Strictly the integers form an integral domain, while the rational numbers, and the real numbers, each form a field. These detailed distinctions are important, but a strict examination is beyond the scope of this book. The reader who wishes to make a more rigorous study of numbers should refer to a more advanced book such as *Basic Mathematics*, R. G. D. Allen, or *Pure Mathematics*, G. H. Hardy; see also the Summary at the beginning of Volume 2 of this book.

The exponential function. The laws of indices

We shall now examine the nature of the function $y = a^x$, where $a, x \in \mathbf{R}^*$ and $a > 0$. When $m, n \in \mathbf{J}^+$ the meaning of a^m is simply $a \times a \times a \times \ldots$ (m factors), and the index laws are easily verified; they are $a^m \times a^n = a^{m+n}$, $a^m/a^n = a^{m-n}$ $(m > n)$, $a^m/a^n = 1/a^{n-m}$ $(m < n)$, $(a^m)^n = a^{mn}$. These laws are of ancient origin and may well have been known to Archimedes, although, of course, he did not have the advantage of an algebraic notation.

When $x \in \mathbf{R}^*$ we shall assume that $y = a^x$ exists and is subject to just two postulates:

$$a^1 = a, \qquad a^m \times a^n = a^{m+n}.$$

We begin by taking $a = 2$ and trying to identify as many ordered pairs of the locus

$$\{(x, y) | y = 2^x\}$$

as we can; we shall then sketch the graph of the function $y = 2^x$.

(i) $2^0 \times 2^1 = 2^{0+1} = 2^1 \Rightarrow 2^0 = 1$.

(ii) $2^{1/2} \times 2^{1/2} = 2^{1/2+1/2} = 2^1 = 2 \Rightarrow 2^{1/2} = \sqrt{(2)}$.

Then $2^{3/2} = 2^{1+1/2} = 2^1 \times 2^{1/2} = 2\sqrt{(2)}$, and so on.

(iii) $2^{-x} \times 2^x = 2^{-x+x} = 2^0 = 1 \Rightarrow 2^{-x} = 1/2^x$.

Then $2^{-1} = 1/2$, $2^{-1/2} = 1/\sqrt{(2)} = \frac{1}{2}\sqrt{(2)}$, and so on.

We now have enough information to evaluate 2^x in the domain $-3 \leqslant x \leqslant 3$ at intervals of 0·5. For example,

$$2^{-2\cdot5} = 2^{-3+0\cdot5} = 2^{-3} \times \sqrt{(2)} \approx 0\cdot18.$$

With a computer it would be possible to evaluate 2^x for $x = \frac{1}{2}, \frac{1}{4}, \frac{1}{8}, \ldots$, and then, expressing x in the binary scale, 2^x could be found by multiplication.† For example, 0·75 becomes 0·11 in the binary scale, and so $2^{0\cdot75} = (\sqrt{2})(\sqrt[4]{2})$.

† Newton's method for evaluating square roots is given in Chapter 1, p. 7.

The table of values and the graph of $y = 2^x$ for $-3 \leqslant x \leqslant 3$ are given below, in Fig. 5.1.

Table of values of $y = 2^x$

x	y
−3·0	0·12(5)
−2·5	0·18
−2·0	0·25
−1·5	0·35
−1·0	0·50
−0·5	0·71
0	1·00
0·5	1·41
1·0	2·00
1·5	2·83
2·0	4·00
2·5	5·66
3·0	8·00

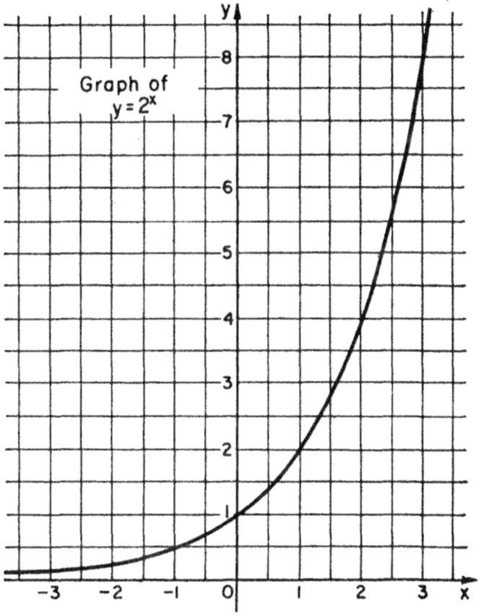

Fig. 5.1

The function a^x, $a > 0$

The postulates are, for $m, n \in \mathbf{R}^*$,

1. $a^1 = a$,
2. $a^m \times a^n = a^{m+n}$.

(i) $a^0 \times a^1 = a^{0+1} = a^1 \Rightarrow a^0 = 1$.

(ii) $a^{-m} \times a^m = a^{-m+m} = a^0 = 1 \Rightarrow a^{-m} = 1/a^m$.

(iii) For $m = p/q$, where $p, q \in \mathbf{J}^+$,

$$a^m \times a^m \times \ldots (q \text{ factors}) = a^{mq} = a^p,$$
$$(a^m)^q = a^p \Rightarrow a^{p/q} = \sqrt[q]{(a^p)}.$$

In particular $a^{1/q} = \sqrt[q]{(a)}$, and furthermore
$$a^{-p/q} = 1/a^{p/q} = 1/\{\sqrt[q]{(a^p)}\}.$$

(iv) If $m, n \in \mathbf{R}$, express m, n with a common denominator t; say $m = p/t$, $n = q/t$, where $p, q, t \in \mathbf{J}^+$.

$$\begin{aligned}(a^{mn})^t &= a^{mnt}, \text{ from (iii)},\\ &= a^{pq/t},\\ &= (a^{p/t})^q, \text{ from (iii)},\\ &= b^q, \text{ where } b = a^{p/t}.\end{aligned}$$

But

$$\begin{aligned}(b^{q/t})^t &= b^q, \text{ from (iii), and so}\\ (a^{mn})^t &= b^q = (b^n)^t.\end{aligned}$$

Hence

$$a^{mn} = b^n = (a^m)^n.$$

With the aid of (ii) this result can now be extended to the case where $m < 0$.

The concept of the function a^x has now been extended over a wider domain $x \in \mathbf{R}$. Figure 5.2 shows the general shape of the graphs of $y = a^x$ when $a > 1$ (continuous line) and $a < 1$ (dotted line). The smooth appearance of the graphs encourages an attempt to extend the domain still further, to $x \in \mathbf{R}^*$.

We need consider only $0 < x < 1$, since $a^{k+x} = a^k \times a^x$ and, if $k \in \mathbf{J}$, the results can be extended over $\mathbf{R}^*$.

Divide the interval $0 \leqslant x \leqslant 1$ into 2^n equal parts and suppose $m/2^n < x < (m+1)/2^n$, noting that x cannot coincide with a point of division. The length of this interval is $1/2^n$ and so we have a set of nested intervals defining x. At the same time, when $a > 1$, y lies in the interval $a^{m/2^n} < y < a^{(m+1)/2^n}$, and again a set of nested intervals defines $y = a^x$.

To sum up, the function $y = a^x$ exists for $x \in \mathbf{R}^*$ and can be

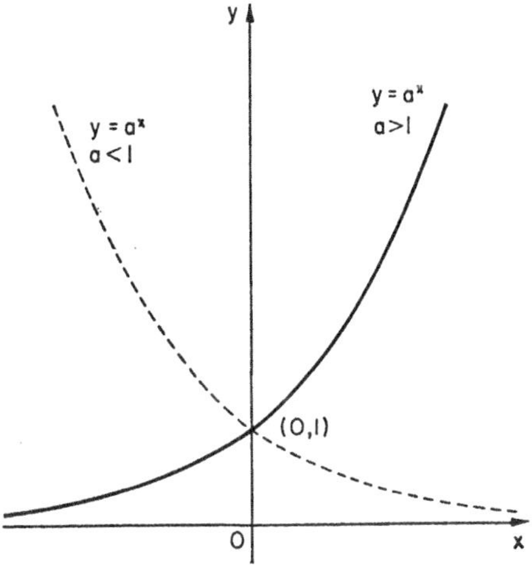

Fig. 5.2

evaluated to any degree of accuracy by methods similar to those used to evaluate 2^x.

It must be noted that, in order to retain its status as a function, a^x must be single valued. In cases like $a^{1/2} = \sqrt{(a)}$, the positive root only can be taken.

EXAMPLE. *Simplify* (*i*) $8^{5/6}/4^{1/4}$, (*ii*) $8^{5/6} - 4^{1/4}$, (*iii*) $64^{-2/3}$.

(i) $8 = 2^3 \Rightarrow 8^{5/6} = (2^3)^{5/6} = 2^{5/2}$.
$4 = 2^2 \Rightarrow 4^{1/4} = (2^2)^{1/4} = 2^{1/2}$.
Hence $8^{5/6}/4^{1/4} = 2^{5/2}/2^{1/2} = 2^2 = 4$.

(ii) $8^{5/6} - 4^{1/4} = 2^{5/2} - 2^{1/2} = 2^{1/2}(4-1) = 3\sqrt{2}$.

(iii) $64 = 2^6 \Rightarrow 64^{-2/3} = (2^6)^{-2/3} = 2^{-4} = 1/16$.

EXERCISE 5.1

In nos. 1–8, *use the laws of indices to simplify:*

1. $2^{1/2} \times 2^{3/2}$. **2.** $(3^2)^{1/2}$. **3.** $3^{5/4}/3^{1/4}$. **4.** $(2^{1/2})^4$. **5.** $3^{5/2} \times 3^{3/2}$.
6. $2^{3/4}/2^{-1/4}$. **7.** $3^{-4} \times 9^3$. **8.** $9^{3/4}/3^{1/2}$.

Evaluate nos. 9–15:

9. 2^{-3}. **10.** 3^{-2}. **11.** $81^{3/4}$. **12.** $16^{-3/2}$. **13.** $27^{4/3}$. **14.** $9^{3/2} \times 27^{-2/3}$.
15. $4^{1/3} \times 6^{1/2} \times 864^{1/6}$.

Evaluate nos. 16–22 *when* $x = 64$:

16. $x^{-1/2}$. **17.** $x^0/x^{1/6}$. **18.** $4x^{1/3}$. **19.** $(4x)^{-1/4}$. **20.** $8x^{1/3}/(16x)^0$.
21. $x^{2/3} \times x^{-3/2} \times x^{1/2}$. **22.** $x^{1/3} - x^{1/2} + 2x^{1/6}$.

Simplify nos. 23–29, *leaving the answer in surd form*:

23. $2^{1/2} + 2^{3/2}$. **24.** $4^{2/3} - 4^{1/3}$. **25.** $5^{-1/2} \times 5^{3/4}$. **26.** $4^{1/3}/2^{1/3}$.
27. $8^{1/2} - 4^{1/4}$. **28.** $9^{3/4} + 3^{1/2} - 3^{-1/2}$. **29.** $8^{1/2} \times 2^2/32^{1/2}$.

Solve, by inspection, nos. 30–35:

30. $x^{1/2} = 3$. **31.** $x^{-1/3} = 2$. **32.** $8^x = 2$. **33.** $7^x = 1$. **34.** $4^x = \frac{1}{2}$.
35. $4^x \times 2^{1-x} = 8$.

36. Solve the equation $x^{1/3} - 3x^{-1/3} = 2$.

37. Evaluate $\dfrac{\sqrt{(x^2+1)}}{x+\sqrt{(x^2+1)}}$ when $x = \frac{1}{2}(a^{1/2} - a^{-1/2})$.

38. Find, by inspection, the value of x when $2^x \times 3^{2x} = 324$.

Logarithms—an inverse function

Figure 5.3 shows the graph of $y = a^x$, $a > 1$. The conditions for an inverse function are satisfied, since there is only one value of x for a given value of y.

This inverse function is so important that it is given a name:

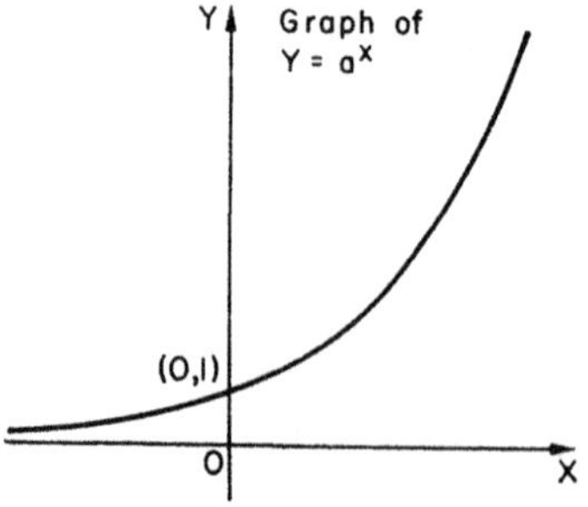

Fig. 5.3

we say that x is the *logarithm* of y, to the *base* a. Thus, if $y = a^x$, then $x = \log_a y$.

The graph of $y = \log_a x$ is shown in Fig. 5.4. It is, of course, the same graph as that in Fig. 5.3, with the independent and dependent variables interchanged.

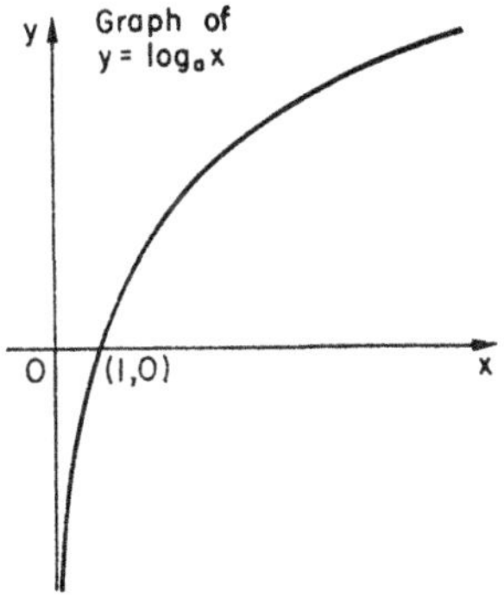

Fig. 5.4

DEFINITION. *If* $x = a^y$, *then* $y = \log_a x$; y *is the logarithm of* x *to the base* a.

The domain of the function $y = \log_a x$ is $0 < x$ and the range is all possible real values.

The laws of logarithms

The laws of indices that are most frequently used in computation are (i) $a^m \times a^n = a^{m+n}$, (ii) $a^m/a^n = a^{m-n}$, (ii) $(a^m)^n = a^{mn}$, although, since $a^{-n} = 1/a^n$, (i) and (ii) are different forms of the same law.

Suppose $a^m = P$, $a^n = Q$, so that $m = \log_a P$, $n = \log_a Q$. But $P \times Q = a^m \times a^n = a^{m+n}$, so that $m + n = \log_a PQ$; this gives

$$\log_a PQ = \log_a P + \log_a Q. \qquad \text{(i)}$$

Similarly $P/Q = a^m/a^n = a^{m-n}$, giving $m - n = \log_a P/Q$.

Hence

$$\log_a P/Q = \log_a P - \log_a Q. \qquad \text{(ii)}$$

Again, $P^k = (a^m)^k = a^{km}$, so that $km = \log_a P^k$; or

$$\log_a P^k = k \log_a P. \qquad \text{(iii)}$$

There is one more important law of logarithms, arising from the necessity to change the base. Suppose, for example, that $\log_a P$ is known and $\log_b P$ is required. Let $t = \log_a P$, so that

$$P = a^t.$$

Hence

$$\log_b P = \log_b a^t = t \log_b a,$$

and so

$$\log_b P = \log_a P \times \log_b a. \qquad \text{(iv)}$$

Suppose that, in (iv), we put $P = b$; then $P = b^1$, so $\log_b P = 1$. Rule (iv) becomes

$$1 = \log_a b \times \log_b a, \quad \text{or } \log_b a = 1/\log_a b.$$

An alternative form of (iv) is $\log_b P = \log_a P/\log_a b$.

An important special case of (iii) concerns $\sqrt[n]{(N)}$, where $n \in \mathbf{J}^+$. Since $\sqrt[n]{(N)} = N^{1/n}$,

$$\log_a \sqrt[n]{(N)} = \frac{1}{n} \log_a N.$$

Logarithms to base 10

In computation where logarithms are used, the base 10 is almost always the most convenient, because the decimal system automatically causes numbers to be expressed as the sum of powers of 10. In the remainder of this chapter it will be assumed that 10 is the base unless another base is shown.

We have $\log 10^n = n$, since $\log 10 = 1$ $(10 = 10^1)$. Important special cases are

$$\log \tfrac{1}{100} = -2, \ \log \tfrac{1}{10} = -1, \ \log 1 = 0, \ \log 10 = 1, \ \log 100 = 2.$$

Books of logarithms (base 10) normally show only the logarithms of numbers between 1 and 10 (standard form), and the familiar *bar notation* is used for the logarithms of numbers less than 1 (unity). For example, if we are given that log 2 = 0·30103,

$$\log 0{\cdot}02 = \log (2/10^2) \quad = \log 2 - 2 = \bar{2}{\cdot}30103,$$

$$\log 2000 = \log (2 \times 10^3) = \log 2 + 3 = 3{\cdot}30103.$$

EXAMPLE. *Evaluate log* 3·9 + 2 *log* 2 – *log* 24 – *log* 6·5.

Since 2 log 2 = log 2^2 = log 4, the given expression is equal to

$$\log \frac{3{\cdot}9 \times 4}{24 \times 6{\cdot}5}$$

$$= \log \tfrac{1}{10} \quad = -1.$$

An alternative method:

$$3{\cdot}9 = \frac{39}{10} = \frac{3 \times 13}{10}, \qquad 24 = 2^3 \times 3, \qquad 6{\cdot}5 = \frac{5 \times 13}{10}.$$

Hence the given expression is equal to

$$(\log 3 + \log 13 - \log 10) + 2 \log 2 - 3 \log 2 - \log 3$$
$$- (\log 5 + \log 13 - \log 10)$$
$$= -\log 2 - \log 5 = -\log 10 = -1.$$

EXAMPLE. *Calculate, to* 3 *sig. fig.*, $\log_5 8$.

We have $\log_5 8 = \log_{10} 8/\log_{10} 5$.

$\log_{10} 5 = 1 - \log_{10} 2 = 0{\cdot}6990,$

$\log_{10} 8 = 0{\cdot}9031.$

Hence

$\log_5 8 = 0{\cdot}9031/0{\cdot}6990$

$= 1{\cdot}29$, correct to 3 sig. fig.

No.	Log.
0·9031	$\bar{1}$·9557
0·6990	$\bar{1}$·8445
1·292	0·1112

EXAMPLE. *Solve the equation* $5^{11-3x} = 2^x$.

$$(11 - 3x) \log 5 = x \log 2,$$
$$x(\log 2 + 3 \log 5) = 11 \log 5,$$
$$x = 11 \log 5/(\log 2 + 3 \log 5)$$
$$= 7{\cdot}689/2{\cdot}398.$$

Hence $x = 3{\cdot}21$, correct to 3 sig. fig.

No.	Log
5	0·6990
2	0·3010
7·689	0·8859
2·398	0·3798
3·207	0·5061

EXERCISE 5.2

In this Exercise all logarithms are to base 10. *No tables are to be used.*

1. Given log 2 = 0·3010, log 3 = 0·4771, find the logarithms to 3 places of decimals of

(i) 4, (ii) 5, (iii) 6, (iv) 8, (v) 9, (vi) 1·2, (vii) 1·5, (viii) 1·6, (ix) 1·8.

In nos. 2–10, *find the value of:*

2. $\log 5 + \log 8 - 2 \log 2$. **3.** $\log 15 + \log 20 - \log 3$.

4. $\log \sqrt{(8)} + \log \sqrt{(24)} - \log \sqrt{(192)}$. **5.** $\log \frac{4}{3} + \log \frac{5}{2} - \log 3\frac{1}{3}$.

6. $\log \sqrt{(\frac{9}{8})} + \log \sqrt{(\frac{2}{3})} - \log \sqrt{3} - \log 5$.

7. $-\log \sqrt{(7)} + \log \sqrt{(175)} - \log 0{\cdot}5$.

8. $\log 8/\log 2$. **9.** $\log 3/\log 9$. **10.** $\log 125/\log 5$.

EXERCISE 5.3

In this Exercise the logarithms, except where otherwise stated, are to base 10. *Give numerical answers correct to* 3 *sig. fig. when not exact.*

1. Evaluate (i) $3 \log_3 10$, (ii) $\log_e 0{\cdot}3$, when $\log_{10} e = 0{\cdot}4343$.

2. Solve the equations:

(i) $2^{3x} \times 3^{4x-1} \times 4^{x-3} = 1$; (ii) $3^{2x} \times 2^{-x} = (2{\cdot}67)^{1/x}$;

(iii) $3 \log x + 5 \log 5 - 7 \log 2 = 5$.

3. If $\log y = 2 - \log x^{2/3}$, express y as a function of x without logarithms. Find y when $x = 8$ without using tables.

Logarithms to any base

The statements $y = a^x$ and $x = \log_a y$ can be written in one single statement, $y = a^{\log_a y}$. In words, if a number y is expressed as a power of a, the index is the logarithm of y to the base a.

EXAMPLE. *Find (i) $\log_2 16$, (ii) $\log_{1/3} 9$.*

(i) $16 = 2^4 \Leftrightarrow \log_2 16 = 4$.

(ii) $9 = (\frac{1}{3})^{-2} \Leftrightarrow \log_{1/3} 9 = -2$.

EXERCISE 5.4

Evaluate nos. 1–12 *without using tables:*

1. $\log_2 8$. **2.** $\log_3 9$. **3.** $\log_5 \sqrt{(125)}$. **4.** $\log_4 (\frac{1}{16})$. **5.** $\log_2 (\frac{1}{32})$.

6. $\log_3 (\frac{1}{243})$. **7.** $\log_{1/2} (\frac{1}{4})$. **8.** $\log_{1/3} 27$. **9.** $\log_{1/4} (64)$.

10. Prove that $\log_a x + \log_a (1/x) = 0$.

11. Prove that $\log_a b \times \log_b c \times \log_c a = 1$.

12. Prove that $\log_{1/a} a^n = \log_{1/b} b^n$.

Napierian logarithms

The first man in Great Britain to invent a device for calculations which later developed into logarithms was Baron Napier of Merchiston (1550–1617). He approached the problem by comparing an arithmetic series, such as

$$0 + 1 + 2 + 3 + 4 + 5 + 6 + \ldots$$

with a geometric series, such as

$$1 + 2 + 4 + 8 + 16 + 32 + 64 + \ldots$$

Instead of multiplying 4×16 we add the corresponding terms of the first series, $2 + 4 = 6$, and read off the term of the second series that corresponds to 6 in the first series.

If Napier's logarithms are compared with the modern system, the base proves to be related to e, a number between 2 and 3 ($e \approx 2{\cdot}718$).† Strangely enough this number e, which is irrational, proves to have quite fundamental importance in the calculus, the subject of later chapters of this book. Modern methods of computing logarithms produce logarithms to base e; these are then converted to base 10 by means of the formula

$$\log_{10} N = \log_e N / \log_e 10.$$

† If N is the base of Napier's logarithms, $\log_N x = 10^7 \log_e (10^7/x)$.

The slide rule

Adding can be done by placing two similarly graduated rulers together, as shown in Fig. 5.5. If the zero of the top scale is placed against the 3 of the bottom scale, the result of adding 3 to any number on the top scale is shown by the number immediately below it on the bottom scale.

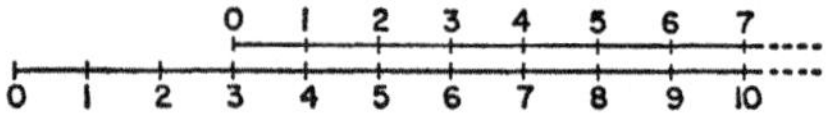

Fig. 5.5

In a slide rule, which is a highly portable form of computer, the rulers are graduated in lengths that are proportional to logarithms, so that the rulers add logarithms, not numbers. A highly simplified version is shown in Fig. 5.6. Although the lengths on the rulers are proportional to logarithms, they are marked with the corresponding numbers. Thus the origin is marked 1 (unity), since $\log 1 = 0$. Figure 5.6 shows the rulers set for multiplication by 1·5. For example $4 \times 1{\cdot}5 = 6$; the 4 on the top scale is immediately above the 6 on the lower scale.

If the distance between the graduations for 1 and 10 is k cm,

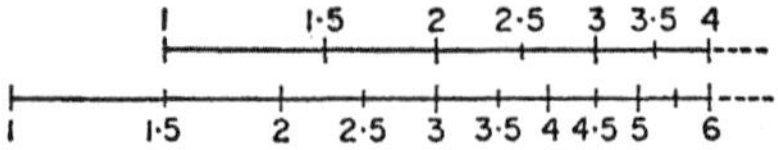

Fig. 5.6

$k = c(\log 10 - \log 1)$, or $c = k$. If the distance between the graduations for 1 and N is x cm,

$$x = c(\log N - \log 1) = c \log N$$

and so

$$x = k \log N.$$

For example, the distance from the first mark (unity) to the mark for 2 is $k \log 2$, and the distance to the mark for 3 is $k \log 3$. The two rulers are, of course, graduated in exactly the same way.

Suppose the unit mark on the upper scale is placed over the A mark on the lower scale, and the reading below the B mark on the upper scale is C on the lower scale. Then

$$k \log C = k \log A + k \log B,$$

or

$$C = AB.$$

Division is done by using the rules for subtraction. Suppose the mark for P on the upper scale is placed over the Q mark on the lower scale ($P < Q$), and the unit mark on the upper scale is above the R mark on the lower scale. Then

$$k \log R = k \log Q - k \log P,$$

or

$$R = Q/P.$$

An actual straight slide rule is much more complicated than this. A strip, called the slide, moves between two fixed scales, as in Fig. 5.7. There are four scales: A is the upper fixed scale and is graduated in the same way as B on the upper edge of the slide; C on the lower edge of the slide is the same as D on the lower fixed scale. A cursor carries a very fine line across all four scales to increase the accuracy of the readings. The A and B scales give the logarithms of numbers from 1 to 100, while the C and D scales, which are of the same length, give the logarithms of numbers from 1 to 10.

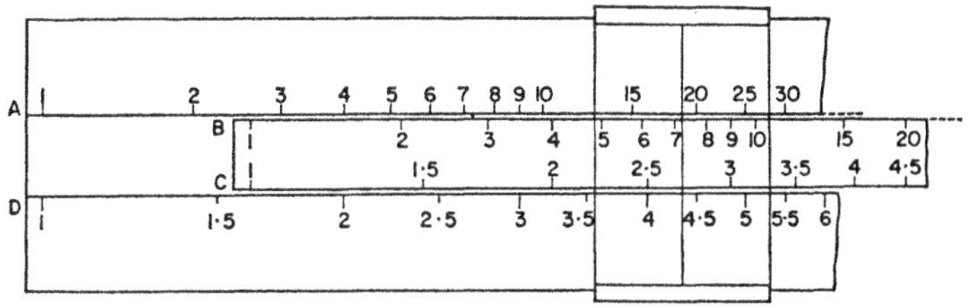

Fig. 5.7

For A and B scales, if x cm measures $\log N$, $x = k \log N$; for the C and D scales, $x = 2k \log N$. Hence, if the cursor shows that a on scale A corresponds to b on scale D, $a = b^2$ or $b = \sqrt{a}$. The four scales therefore allow multiplication, division, squares and square roots. A good quality slide rule about 30 cm long gives readings accurate to 3 significant figures.

Many other operations can be carried out on a good slide rule, and special handbooks are provided by the manufacturers to explain how to use them. A model slide rule can easily be made on stiff cardboard, especially if logarithmic graph paper is available.

EXAMPLE. *On a scale 30 cm long find the distance between the graduations for 1·8 and 5·5 (i) on the A and B scales, (ii) on the C and D scales.*

On the A and B scales, $c \log 100 = 30$, so that $c = 15$.

The distance between the 1·8 and 5·5 graduations is therefore

$$\begin{aligned} & 15\,(\log 5{\cdot}5 - \log 1{\cdot}8) \text{ cm} \\ = {} & 15\,(0{\cdot}7404 - 0{\cdot}2553) \text{ cm} \\ = {} & 7{\cdot}28 \text{ cm (to 3 sig. fig.).} \end{aligned}$$

On the C and D scales the distance will be just double, that is

$$2 \times 7{\cdot}2765 = 14{\cdot}6 \text{ cm (to 3 sig. fig.).}$$

EXERCISE 5.5

1. Use a slide rule to evaluate the following; if no slide rule is available explain the setting of the scales, with diagrams:

(i) $2{\cdot}4 \times 3$, (ii) $8{\cdot}4 \times 3$, (iii) $5{\cdot}4/2{\cdot}8$, (iv) $5{\cdot}4/8{\cdot}5$, (v) $(3{\cdot}5)^2 \times 4{\cdot}6$, (vi) $\sqrt{(15)}/3{\cdot}5$, (vii) $3{\cdot}4 \times 2{\cdot}8/2{\cdot}2$, (viii) $4{\cdot}5 \times 8{\cdot}5/9{\cdot}6$, (ix) $(5{\cdot}6/3{\cdot}7)^2 \times 17{\cdot}5$.

2. The length of the C and D scales on a slide rule is 12 in. Find the distance between the graduations for:

(i) 1·85 and 3·52, (ii) 1·08 and 9·45.

Write down the distances between the same graduations on the A and B scales.

3. The A and B scales on a slide rule are 20 cm long. Find the distances between the graduations for:

(i) 7·8 and 8·1, (ii) 78 and 81.

4. If a slide rule one metre long were graduated from 1 to 1000, find what the distances would be between the graduations for:

(i) 5 and 350, (ii) 8 and 84, (iii) 12 and 960.

5. The distance between the graduations for 2 and 3 on a slide rule is 2·5 cm. What are the readings at points on either side of the graduation for 3, and one centimetre away from it? If the scale is graduated from 1 to 100, what is its length?

Graphical solution of equations

If the graphs of $y = 4^x$ and $y = 2(2x + 3)$ and drawn on the same axes and with the same scales they intersect at two points, A and B. The coordinates (x, y) of these two points satisfy the equations of both graphs simultaneously. Hence the x-coordinates of A and B are solutions of the equation

$$4^x = 2(2x + 3).$$

This gives a practical method of solving the equation, provided solutions can be found within the necessary limits of accuracy.

When

$$x = -2, \quad 4^x = 0{\cdot}0625, \quad 2(2x + 3) = -2,$$
$$x = 0, \quad 4^x = 1, \quad 2(2x + 3) = 6,$$
$$x = 2, \quad 4^x = 16, \quad 2(2x + 3) = 14.$$

The roots therefore lie between -2 and 0, and between 0 and 2, and we find a first approximation by drawing the two graphs, as in Fig. 5.8.

x	$-2{\cdot}0$	$-1{\cdot}5$	$-1{\cdot}0$	$-0{\cdot}5$	0	0·5	1·0	1·5	2·0
4^x	0·06	0·12(5)	0·25	0·50	1·00	2·00	4·00	8·00	16·00
$2(2x + 3)$		0			6				14

A first approximation to the roots is

$$x = -1{\cdot}45 \text{ or } 1{\cdot}85.$$

The method for finding a second and third approximation will be shown only for the point A, where $x \approx 1{\cdot}85$. It will be

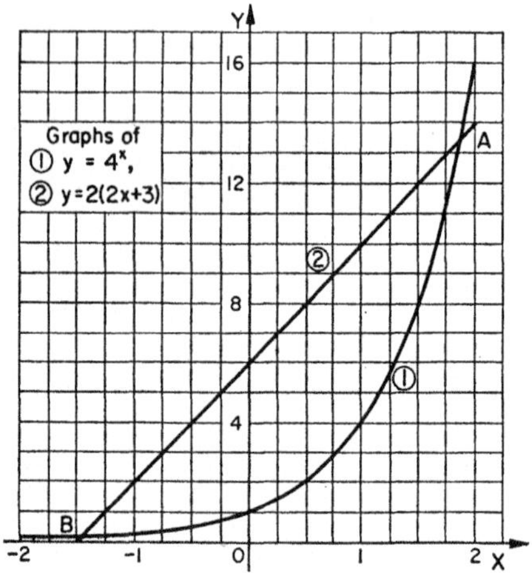

Fig. 5.8

seen that, over the narrow intervals involved, the graph of $y = 4^x$ is treated as a straight line; also the use of a false origin makes it possible to employ a larger scale.

If $a = 4^{1\cdot8}$, $\log_{10} a = 1\cdot8 \log_{10} 4 = 1\cdot8 \times 0\cdot6021$, $\log_{10} a = 1\cdot0838 \Rightarrow a = 12\cdot13$.

Similarly $4^{1\cdot9} = 13\cdot93$.

The drawing for the second approximation is given in Fig. 5.9.

x	1·8	1·9
4^x	12·1	13·9
$2(2x + 3)$	13·2	13·6

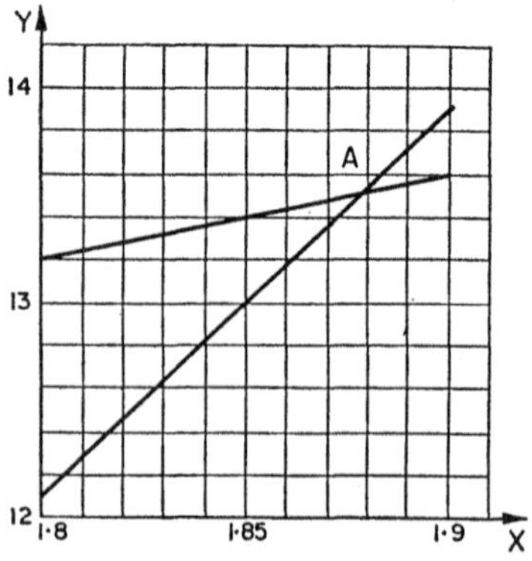

Fig. 5.9

The second approximation is $x \approx 1{\cdot}88$. A third approximation could now be performed in the same way, between 1·87 and 1·89, and that is about the limit of accuracy possible with four-figure tables.

EXERCISE 5.6

1. Draw the graph of $y = 9^x$ for $-1 \leqslant x \leqslant 1$. Use it with another graph to find a second approximation for the (positive) root of:
(i) $9^x = 5 - 4x$, (ii) $9^x = 4x + 2$.

2. Draw the graph of $y = 2^x$ for $-2 \leqslant x \leqslant 3$. Find a second approximation for the (positive) root of
(i) $2^x = \frac{3}{2}(x + 2)$, (ii) $x \, . \, 2^x = 1$.

3. Draw the graph of 3^x for $-2 \leqslant x \leqslant 2$. Find a second approximation for the root of (i) $3^x = 4x$ (the smaller root), (ii) $x^2 \, . \, 3^x = 1$.

EXERCISE 5.7
(*Miscellaneous*)

No tables are to be used in nos. 1–25. *All logarithms are to base* 10 *unless otherwise shown.*

Simplify nos. 1–11*:*

1. $16^{3/4}/64^{-1/3}$. **2.** $32^{-2/5} \times (3\frac{3}{8})^{-1/3}$. **3.** $8^0 \times 16^{1/4}/81^{-1/4}$. **4.** $(4x^{-4})^{-5/2}$.
5. $(9x^{-2})^{-1/2}$. **6.** $(3x^{-1/3})^3$. **7.** $\log 16 - \log 2 - \log 20$.
8. $2 \log x - \log \sqrt{(x)} + 3 \log (1/\sqrt{x})$. **9.** $\log 27/\log 9$.
10. $(\log \sqrt{2} + \log \sqrt{3})/\log 6$. **11.** $(\log 9 - \log 4)/(\log 3 - \log 2)$.

Solve the equations in nos. 12–22*:*

12. $3^x = 729$. **13.** $3^{2x} - 10.3^x + 9 = 0$. **14.** $2^{2x} - 3 \, . \, 2^{x+1} = -8$.
15. $a^x + a^{-x} = 2$. **16.** $\log x = \log 15 - \log 9$. **17.** $\log x = 1 + \log 5$.
18. $\log 25 - \log x = (\log 8)/(\log 2)$. **19.** $\log (x + 1) - \log x = 2$.
20. $2^x - 3^y = -5$, $2^{x-1} + 3^{y+1} = 29$.
21. $2 \, . \, 3^x - 4 \, . \, 7^y = 2$, $3^{x-1} + 7^{y+1} = 8$. **22.** $\log_x 2 = \frac{1}{2}$.

Evaluate nos. 23–32*:*

23. $\log_4 256$. **24.** $\log_2 (\frac{1}{4})$. **25.** $\log_4 32$. **26.** $\sqrt[5]{(0{\cdot}8427)}$.
27. $(0{\cdot}5)^{-1/4}$. **28.** $(0{\cdot}38)^{2{\cdot}5}$. **29.** e^{-x^2}, when $x = 1{\cdot}1$, $\log e = 0{\cdot}4343$.
30. $\log_2 5$. **31.** $\log_3 7$. **32.** $\log_4 0{\cdot}4$.

Solve the equations in nos. 33–35*:*

33. $5^x = 4{\cdot}3$. **34.** $(1{\cdot}015)^x = 2$. **35.** $(0{\cdot}3)^x = 0{\cdot}56$.

36. If $pv^n = c$, find n when $p = 150$, $v = 45$, $c = 3{\cdot}53 \times 10^5$.

37. Sketch the graph of 10^x and hence find the number of roots of the equations:

(i) $10^x = 4 - x$, (ii) $10^x = 2x + 4$.

In each case state between what integral values x must lie.

Draw the graphs accurately and find the solutions correct to one place of decimals.

38. Repeat no. 37 with the graph of $\log x$ and the equations:

(i) $x \log x = 0{\cdot}2$, (ii) $\log x = 0{\cdot}3x - 0{\cdot}5$.

Write down also, to one sig. fig., the solutions of $\log x = 1{\cdot}1x - 10$.

39. Use a graphical method to find a solution of the equation

$$2^{x/90} = 2/x.$$

correct to 3 sig. figs.

40. In 1961 it was said that the population of the world was increasing at the rate of 2 per cent per annum of what it was at the beginning of the year. If this rate of increase remained constant, what would be the first year in which the population was double that at the beginning of 1961?

CHAPTER 6

THE STRAIGHT LINE

Projections—the distance formula

Suppose P and Q are any two points in the plane XOY, and PM, QN are the perpendiculars from P, Q to OX. PR is the perpendicular from P to QN, or to QN produced, or to NQ produced.

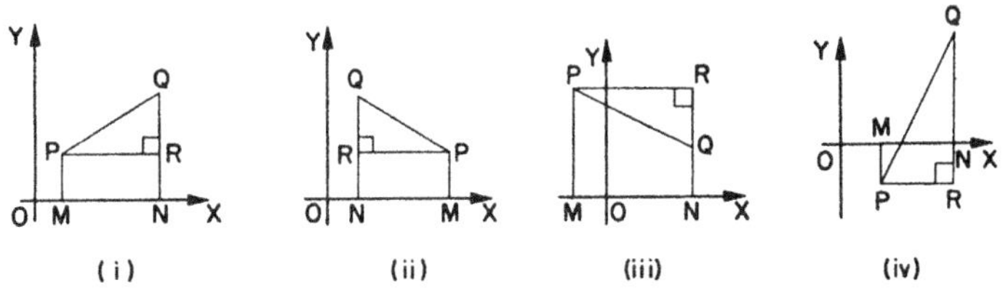

Fig. 6.1

Figure 6.1 shows four arrangements of P and Q in the four quadrants; in each case $\Delta\, PQR$ has a right angle at R and $\overrightarrow{PR} = \overrightarrow{MN}$.

$\overrightarrow{MN}$ is called the *projection* of $\overrightarrow{PQ}$ on OX. If P is (x_1, y_1) and Q is (x_2, y_2), then $\overrightarrow{OM} = x_1$, $\overrightarrow{ON} = x_2$ and $\overrightarrow{MN} = x_2 - x_1$.

Ex. 1. Sketch other positions of PQ in relation to the four quadrants and verify in each case that $\overrightarrow{MN} = x_2 - x_1$.

The projection of $\overrightarrow{PQ}$ on $\overrightarrow{OY}$ is similarly defined, and we have:

$$\textbf{projection of } \overrightarrow{PQ} \textbf{ on } OX = x_2 - x_1,$$

$$\textbf{projection of } \overrightarrow{PQ} \textbf{ on } OY = y_2 - y_1.$$

We can also verify from the figures that the projections of PQ on the axes are equal in length to the sides PR, RQ of $\triangle PQR$; in other words

$$PR = |x_2 - x_1|, \qquad RQ = |y_2 - y_1|.$$

Since $P\hat{R}Q$ is a right angle, the theorem of Pythagoras gives $PQ^2 = PR^2 + RQ^2$, and so

$$PQ^2 = (x_2 - x_1)^2 + (y_2 - y_1)^2.$$

This formula provides the first essential of a metrical geometry, that is, of a geometry dealing with measurements, and it presupposes a knowledge of plane geometry at least as far developed as the theorem of Pythagoras. Alternatively we could base a system of geometry on coordinates, and *define* the length of PQ to be $\sqrt{\{(x_2 - x_1)^2 + (y_2 - y_1)^2\}}$.

Ex. 2. Find the lengths of the lines joining the pairs of points (i) (2, 2), (5, 6); (ii) (2, 2), (−2, 5); (iii) (−4, −2), (8, 3); (iv) (−8, 2), (16, −5).

Ex. 3. Prove that (−3, 4), (4, 5), (2, −6) are the vertices of an isosceles triangle.

Ex. 4. Write down the condition for $P(x, y)$ to be equidistant from $A(-3, 4)$ and $B(4, 5)$. Hence find the equation of the perpendicular bisector of AB.

Ex. 5. Write down the equation of the circle with centre (2, 1) and radius 3.

Gradient

The second requirement in a metrical geometry is the measurement of angle, or of direction.

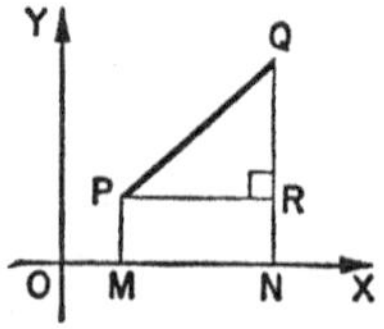

Fig. 6.2

If, in Fig. 6.2, P is (x_1, y_1) and Q is (x_2, y_2), the *gradient* of $\overrightarrow{PQ}$ is defined as

$$\frac{y_2 - y_1}{x_2 - x_1}$$

This can also be described as

$$(\text{projection of } \overrightarrow{PQ} \text{ on } OY)/(\text{projection of } \overrightarrow{PQ} \text{ on } OX)$$

and also as (y-step)/(x-step) in the displacement from P to Q.

In Fig. 6.3 the gradient of PQ is RQ/PR and the gradient of UV is WV/UW. Suppose these gradients are equal; then, since

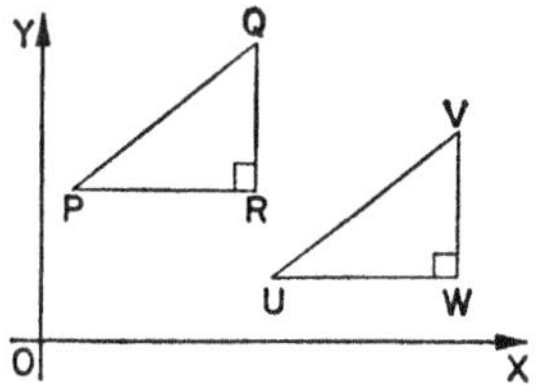

Fig. 6.3

$\hat{R}$ and $\hat{W}$ are right angles, the triangles PRQ and UWV are similar. Hence PQ and UV make equal angles with OX, and so straight lines which have the same gradient are parallel.

Ex. 6. Show that parallel lines have the same gradient.

Ex. 7. Write down the gradients of the lines joining the pairs of points (i) (2, 1), (8, 3); (ii) (−3, 2), (3, −2); (iii) (−3, −1), (2, 2).

Ex. 8. Prove that the line joining (4, 3) and (8, 5) is parallel to the line joining (3, 0) and (5, 1).

Ex. 9. Prove that (−3, 4), (−1, −1), (1, 0) and (−1, 5) are the vertices of a parallelogram by using gradients. Check the result by showing that the diagonals have the same mid-point.

Ex. 10. If A, B, C are the points (1, 5), (−2, 3), (3, −3), prove that the line joining the mid-points of AB, AC is parallel to BC and equal in length to $\frac{1}{2}BC$.

Positive and negative gradients

The experience of the last few exercises will have shown that:

(i) a line which is rising (i.e. on which the value of y is increasing) as we proceed from left to right has a positive gradient;

(ii) a line which is falling as we proceed from left to right has a negative gradient.

Ex. 11. What is the gradient of a line parallel to (i) the x-axis, (ii) the y-axis?

The definition of a gradient on p. 85 would clearly be faulty if it did not produce the same result for any two distinct points on the same straight line. The line joining any two of the points $(-3, -6)$, $(-1, -3)$, $(1, 0)$, $(3, 3)$ has a gradient $\frac{3}{2}$, and so these four points are *collinear*, that is, lie on one straight line.

Every linear equation in x and y represents a straight line.

Suppose $P_1(x_1, y_1)$ is a fixed point on the line $ax + by + c = 0$, and $P(x', y')$ is a variable point on the line. Then the following equations are simultaneously true:

$$ax' + by' + c = 0,$$

$$ax_1 + by_1 + c = 0.$$

Subtracting:

$$a(x' - x_1) + b(y' - y_1) = 0,$$

and so

$$\frac{y' - y_1}{x' - x_1} = -\frac{a}{b}.$$

Hence the gradient of P_1P is constant, and so

$ax + by + c = 0$ is the equation of a straight line with gradient $-a/b$.

In terms of a locus we can say that, if a, b and k are constant, the equation $ax + by = k$ determines the locus

$$\{(x, y) \mid ax + by = k\};$$

the graph of the locus is a straight line with gradient $-a/b$.

If k takes different values we get a *family* of parallel straight lines, each with gradient $-a/b$.

Equation of the straight line joining two points

In Fig. 6.4, A is (2, 3) and B is (5, 4). The gradient of AB is

$$\frac{4-3}{5-2} = \frac{1}{3}.$$

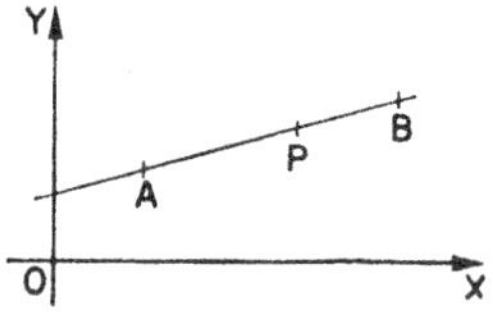

Fig. 6.4

Suppose $P(x, y)$ is any point on the line AB. The gradient of AP is $(y-3)/(x-2)$, and so

$$\frac{y-3}{x-2} = \frac{1}{3},$$

which reduces to $x - 3y + 7 = 0$. Hence every point (x, y) on the line AB satisfies $x - 3y + 7 = 0$.

Conversely, the locus $\{(x, y) \mid x - 3y + 7 = 0\}$ consists of points (x, y) such that $(y-3)/(x-2) = \frac{1}{3}$, since $(x \neq 2)$

$$x - 3y + 7 = 0 \Leftrightarrow \frac{y-3}{x-2} = \frac{1}{3},$$

that is, the locus is the line AB.

We say that the equation of the line joining (2, 3) and (5, 4) is $x - 3y + 7 = 0$.

Note that, if the gradient of a line is $\frac{1}{3}$, its equation is of the form $x - 3y = k$, since the gradient of $ax + by + c = 0$ is $-a/b$. If the line also passes through (5, 4) we have $x - 3y = 5 - 3(4) = -7$, giving $x - 3y + 7 = 0$ as before.

The equation of the straight line with gradient m, passing through (x_1, y_1) is $(y - y_1)/(x - x_1) = m$, or

$$mx - y = (mx_1 - y_1).$$

The equation $y = mx + c$

Suppose $y = mx + c$. When $x = 0, y = c; x = 1, y = m + c$. Hence the line has a gradient m, and cuts the y-axis at a distance c from the origin. (See Fig. 6.5.)

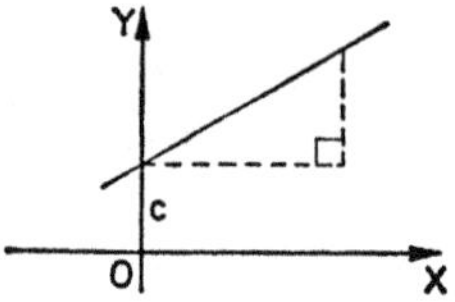

Fig. 6.5

The intercept form

Suppose a line cuts OX at $(a, 0)$ and OY at $(0, b)$. (See Fig. 6.6.) We say that the line cuts off intercepts a and b on the axes.

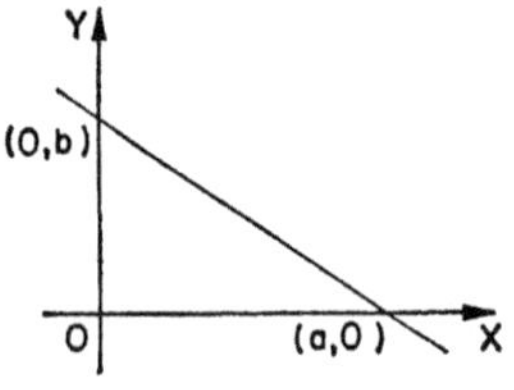

Fig. 6.6

If the equation of the line is $lx + my + n = 0$, we have

$$la + n = 0, \qquad mb + n = 0,$$

and so

$$la = mb = -n,$$

or

$$l : m : n = 1/a : 1/b : -1.$$

Hence the line which cuts off intercepts a and b on the axes is

$$\frac{x}{a} + \frac{y}{b} = 1.$$

Ex. 12. Write down the gradients of the lines
(i) $2x - y + 3 = 0$; (ii) $2x + 3y = 4$; (iii) $2y + 3 = x$; (iv) $2y = 3$; (v) $2y = 3x$; (vi) $2x = 3$.

Ex. 13. Find the equations of the lines joining the pairs of points
(i) (5, 6), (3, 2); (ii) (−2, 3), (4, −1); (iii) (−3, 2), (4, 2); (iv) (3, −2), (3, 4).

Ex. 14. Show that the four lines $y = 2x + 1$, $3x + y = 4$, $2x - y = 7$, $3x + y + 6 = 0$ form a parallelogram.

Ex. 15. Find the equation of the line
(i) through (3, 2) with gradient 3; (ii) through (−2, 1) with gradient $-\frac{4}{3}$; (iii) through (−1, 2) parallel to $2x - 3y = 7$; (iv) through the origin parallel to $2x + 3y = 5$.

Perpendicular lines

Suppose P is (a, b), as in Fig. 6.7. Let the plane rotate through a right angle, carrying OP to OQ, while OX and OY remain fixed. Then Q is $(-b, a)$.

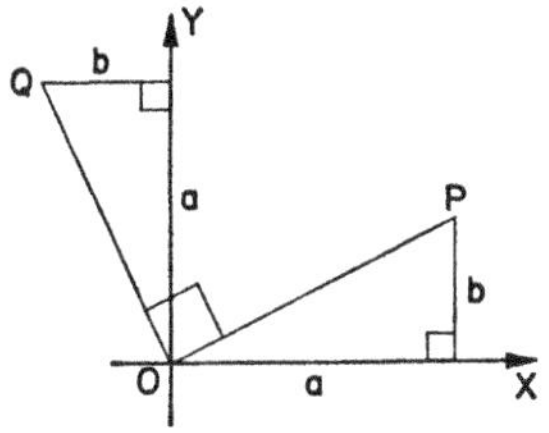

Fig. 6.7

The gradient m_1 of OP is b/a, and the gradient m_2 of OQ is $a/(-b)$, $= -a/b$. Hence $m_1 m_2 = -1$.

Similarly, if P_1 is (x_1, y_1) and P_2 is (x_2, y_2), and the line P_1P_2

is rotated through a right angle about O into the position Q_1Q_2, then Q_1 is $(-y_1, x_1)$ and Q_2 is $(-y_2, x_2)$. The gradient of P_1P_2 is

$$\frac{y_2 - y_1}{x_2 - x_1},$$

and the gradient of Q_1Q_2 is

$$\frac{x_2 - x_1}{-y_2 + y_1} = -\frac{x_2 - x_1}{y_2 - y_1}.$$

Hence, if m_1, m_2 are the gradients of P_1P_2, Q_1Q_2, then

$$m_1m_2 = -1.$$

For example, if the gradient of a given line is $\frac{2}{3}$, the gradient of a perpendicular line is $-\frac{3}{2}$; also the equation of a line perpendicular to $ax + by + c = 0$ is of the form $bx - ay = k$.

EXAMPLE. *Prove that the points* $(1, -3)$, $(-1, 1)$, $(5, 4)$ *form three corners of a rectangle and find the fourth corner and the area of the rectangle.*

The gradient of AB is

$$\frac{1 + 3}{-1 - 1} = -2,$$

and of BC is

$$\frac{4 - 1}{5 + 1} = \frac{1}{2},$$

so AB is $\perp$ BC. (See Fig. 6.8.)

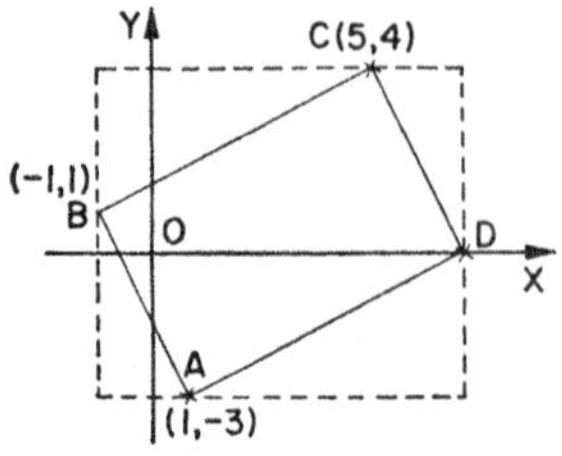

Fig. 6.8

In going from B to C the x-step is 6 and the y-step is 3.

The same is true for going from A to D; since $\overrightarrow{AD} = \overrightarrow{BC}$. Hence D is $(1 + 6, -3 + 3)$, i.e. $(7, 0)$.

The area can be found by subtracting four right-angled triangles from the surrounding rectangle. Hence the area

$$= (7 + 1)(4 + 3) - (6 \times 3) - (4 \times 2) \text{ square units}$$
$$= 30 \text{ square units.}$$

Alternatively,

$$AB = \sqrt{(4^2 + 2^2)} = \sqrt{20},$$
$$BC = \sqrt{(3^2 + 6^2)} = \sqrt{45}.$$

Hence the area

$$= \sqrt{20} \times \sqrt{45} \text{ square units}$$
$$= \sqrt{900} \text{ square units}$$
$$= 30 \text{ square units.}$$

Ex. 16. Find the equation of the line (i) through $(2, 3)$ perpendicular to $3x + y + 2 = 0$; (ii) through $(2, -1)$ perpendicular to $3x - 4y = 7$; (iii) through $(0, 1)$ perpendicular to $y = 3x + 5$; (iv) through the origin perpendicular to $x + y = 0$.

Ex. 17. Show that the lines $3x + 2y = 5$, $2x - 3y + 5 = 0$, $3x + 2y + 7 = 0$, $4x - 6y = 7$ form a rectangle.

Distance of a point from a line

Suppose AB, $ax + by + c = 0$, is a given line and $P_1(x_1, y_1)$ is a given point. (See Fig. 6.9.)

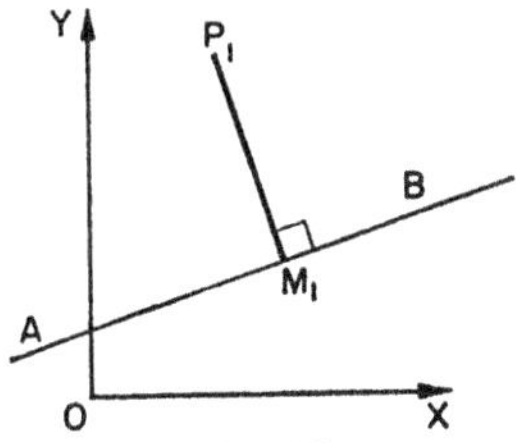

Fig. 6.9

Let M_1 be the foot of the perpendicular from P_1 to AB, and let the coordinates of M_1 be (x_2, y_2).

The gradient of AB is $-a/b$, and so the gradient of P_1M_1 is b/a; then

$$\frac{y_2 - y_1}{x_2 - x_1} = \frac{b}{a}.$$

Suppose $x_2 - x_1 = ma$, $y_2 - y_1 = mb$. Since M_1 lies on AB,

$$ax_2 + by_2 + c = 0.$$

Let

$$ax_1 + by_1 + c = k.$$

Then

$$a(x_2 - x_1) + b(y_2 - y_1) + k = 0,$$

so that

$$m(a^2 + b^2) + k = 0.$$

But

$$\begin{aligned} P_1M_1^2 &= (x_2 - x_1)^2 + (y_2 - y_1)^2 \\ &= m^2(a^2 + b^2) \\ &= \frac{k^2}{(a^2 + b^2)^2} \times (a^2 + b^2). \end{aligned}$$

Hence

$$P_1M_1 = \frac{|k|}{\sqrt{(a^2 + b^2)}} = \frac{|ax_1 + by_1 + c|}{\sqrt{(a^2 + b^2)}}.$$

The perpendicular distance of (x_1, y_1) from $ax + by + c = 0$ is therefore:

$$\frac{|ax_1 + by_1 + c|}{\sqrt{(a^2 + b^2)}}.$$

Ex. 18. Find the distance of
(i) (0, 7) from $4x - 3y + 25 = 0$; (ii) $(-1, 2)$ from $3x + 4y - 1 = 0$; (iii) the origin from $x - y = 2$; (iv) (1, 2) from $5x + 12y - 42 = 0$.

Intersection of two lines

In Fig. 6.10, suppose AB is $4x - 3y + 1 = 0$ and CD is $x + 3y - 11 = 0$. The coordinates of P satisfy both these equations, and so we can find P by solving

$$4x - 3y = -1$$

and

$$x + 3y = 11$$

as simultaneous equations. Thus P is (2, 3).

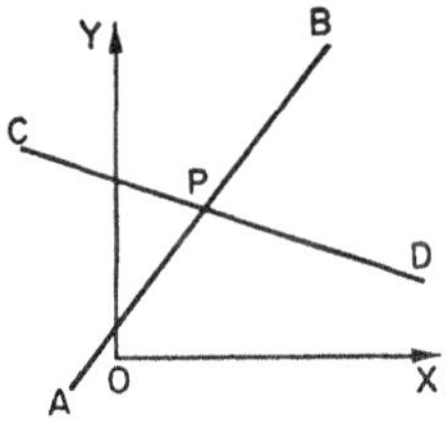

Fig. 6.10

Ex. 19. Find the point of intersection of the pairs of lines (i) $3x + 4y = 1$, $2x + y - 4 = 0$; (ii) $3x - 5y = 10$, $4x + y + 2 = 0$.

Ex. 20. Find the gradient of the line joining the point (1, 2) to the intersection of the lines $x + 3y + 3 = 0$, $3x - y = 11$.

EXAMPLE. *Write down the family of lines parallel to* $4x + 3y = 2$, *and find the two which are at a distance* 4 *units from the given line.*

Since the gradients are the same the family of lines is

$$4x + 3y = k.$$

We want to find a point on one of these lines. Consider the point where the line $4x + 3y = k$ cuts the x-axis; it is $(\frac{1}{4}k, 0)$. This is 4 units from the line $4x + 3y = 2$, and so

$$\frac{k + 0 - 2}{5} = \pm 4,$$

giving $k = 22$ or -18.

The two lines are $4x + 3y = 22$ and $4x + 3y + 18 = 0$.

EXAMPLE. *Prove that the medians of any triangle are concurrent by finding their equations.*

In Fig. 6.11, let A, B, C be the vertices of the triangle. There is no loss of generality in placing the axes in convenient positions. Let BC be the x-axis, and let O, the mid-point of BC, be the origin. We can then let B be $(-a, 0)$, C be $(a, 0)$ and A be (h, k).

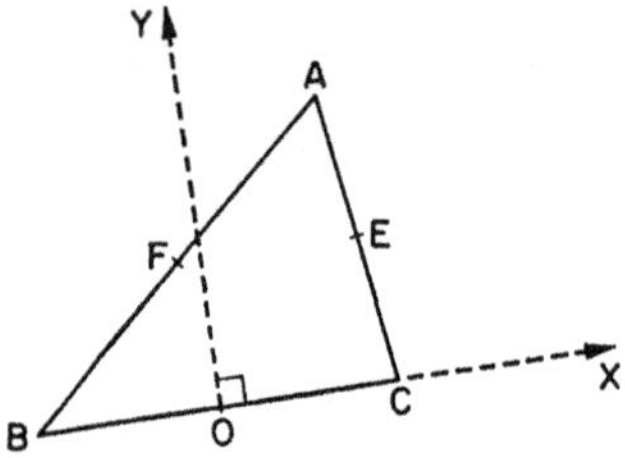

Fig. 6.11

The equation of median AO is

$$\frac{y}{x} = \frac{k}{h}, \text{ or } xk - yh = 0. \tag{i}$$

E, the mid-point of AC is $(\frac{1}{2}(h + a), \frac{1}{2}k)$. The gradient of BE is $k/(h + 3a)$, and it passes through $(-a, 0)$; the equation of BE is

$$kx - (h + 3a)y + ka = 0. \tag{ii}$$

Replacing a by $-a$, the equation of CF is

$$kx - (h - 3a)y - ka = 0. \tag{iii}$$

From (ii) and (iii), subtracting,

$$6ay = 2ka, \text{ or } y = \tfrac{1}{3}k.$$

Substituting in (ii): $kx = \frac{1}{3}k(h + 3a) - ka = \frac{1}{3}hk$, and so BE, CF meet at $(\frac{1}{3}h, \frac{1}{3}k)$.

This point also lies on AO, $kx - hy = 0$, and so the three medians are concurrent. The point of concurrence is the centroid of A, B, C, as defined on p. 42.

Notes.

(i) Axes of coordinates need not be at right angles, provided we do not use gradients or distances. Here OC and OA

could be taken as axes of "oblique coordinates", with A as $(0, b)$. Then E is $(\frac{1}{3}a, \frac{1}{2}b)$ and the equation of BE is $bx - 3ay + ab = 0$.

(ii) This property of the triangle is more easily proved by vectors. Suppose P is any point and G is given by

$$3\overrightarrow{PG} = \overrightarrow{PA} + \overrightarrow{PB} + \overrightarrow{PC}$$

$$= \overrightarrow{PA} + 2\overrightarrow{PO}.$$

Then G lies on AO and $AG{:}GO = 2{:}1$.

Similarly G lies on the other two medians and divides each of them in the ratio 2:1.

EXERCISE 6.1

1. A, B, C, D are the four points (x_1, y_1), (x_2, y_2), (x_3, y_3), (x_4, y_4). P, Q, R, S are the mid-points of AB, BC, CD, DA. Find the coordinates of the mid-points of PR and QS, and show that PR, QS bisect each other.

Show also that the line joining the mid-points of AC and BD passes through the intersection of PR and QS.

2. A point P moves so that the square of its distance from the line $x + y = 1$ is equal to the product of its distances from the lines $3x + y + 2 = 0$ and $x + 3y - 1 = 0$. Show that the coordinates of P satisfy either $2x^2 + 2y^2 - 9x - 15y + 7 = 0$ or $8x^2 + 8y^2 + 20xy - 11x - 5y + 3 = 0$.

3. Find the equation of the straight line which bisects at right angles the line joining $(2, \sqrt{3})$ and $(5, 2\sqrt{3})$.

Show that this line cuts the x-axis at a point which gives with the two previous points the vertices of an equilateral triangle.

4. Find the coordinates of the fourth vertex D of the parallelogram $ABCD$, given that A, B, C are the points $(4, 3)$, $(-1, 2)$, $(2, -1)$.

Find also the coordinates of the point of intersection of the diagonals of this parallelogram.

5. (i) Find the equation of the straight line which makes an intercept of $+\frac{3}{2}$ on the y-axis and passes through $(-1, 2)$.

(ii) The straight line through the point $(4, 5)$ which is at right angles to the line $x + 2y = 3$ cuts it at P. Find the coordinates of the point P.

6. Calculate the area of the triangle formed by the line $2x + 3y = 7$ with the axes.

Find the equation of the line through the origin which divides the triangle into two parts equal in area.

7. The points $(2, 1)$, $(1, 5)$, $(4, 6)$ are the mid-points of the sides of a triangle. Find the equations of the sides.

8. Find the coordinates of the point Q where the perpendicular from P $(0, 4)$ to $2x - 3y = 1$ cuts this line.

If PQ is produced to R so that $PQ = QR$, find the coordinates of R.

Find also the coordinates of the point S on the y-axis which is equidistant from P and R, and also those of the point T such that $PSRT$ is a rhombus.

9. Find the equation of the locus of points equidistant from $(-1, 3)$ and $(2, 4)$.

Find the point which is equidistant from $(-1, 3)$, $(2, 4)$ and $(1, 2)$.

10. The points A, B, C are $(-3, -1)$, $(1, 1)$, $(9, 5)$. Prove that A, B, C are collinear (i) by using gradients; (ii) by showing that $AC = AB + BC$; (iii) by finding the point which divides AC internally in the ratio $1:2$; (iv) by finding the equation of AB.

11. Explain why the equation $(5x - y - 7) + k(3x + 7y - 2) = 0$ represents a line which passes through the intersection of $5x - y - 7 = 0$ and $3x + 7y - 2 = 0$.

Find the equation of the line which passes through the intersection of these two lines and has a gradient $-\frac{4}{3}$.

12. Find the equation of the family of lines through the intersection of $2x + 3y - 7 = 0$ and $5x - 4y + 3 = 0$.

Find that member of the family which is perpendicular to $7x - 3y = 21$.

13. Find the equations of the medians of the triangle whose vertices are $(7, 3)$, $(3, 9)$, $(-3, -1)$, and verify that they all pass through the same point.

14. Find the values of c so that the point $(2, 1)$ is 5 units away from $5x + 12y + c = 0$.

15. Prove that the perpendicular bisectors of the three sides of any triangle are concurrent.

CHAPTER 7

THE LINEAR FUNCTION

Polynomials

An expression of the form

$$a_0x^n + a_1x^{n-1} + a_2x^{n-2} + \ldots + a_{n-1}x + a_n,$$

where the indices n, $(n - 1)$, . . . are all positive integers or zero, is called a polynomial in the variable x. The constants $a_0, a_1, a_2, \ldots a_n$ are called the *coefficients*. The *degree* of the polynomial when $a_0 \neq 0$ is n. The term of highest degree, a_0x^n, is the *leading term*, and a_0 is the *leading coefficient*. a_n is called the *constant* term, or the *absolute* term.

Examples of polynomials are:

(i) 0, the zero polynomial;
(ii) 2, a constant;
(iii) $3x - 4$, a linear polynomial;
(iv) $2x^2 - 3x + 7$, a quadratic polynomial, or a quadratic;
(v) $-x^3 + x^2 - \frac{5}{2}x + \frac{1}{2}$, a cubic polynomial, or a cubic.

Suppose $P(x)$ represents a polynomial. The set $\{(x, y) | y = P(x)\}$ determines a function since there can be only one value of y for each value of x, but the inverse function $x = P^{-1}(y)$ exists only in special cases.

The value of $P(x)$ when $x = a$ is denoted by $P(a)$. For example, if $P(x) = 5x - 3 - x^2$,

$P(-1) = -5 - 3 - 1 = -9, \quad P(0) = -3, \quad P(1) = 5 - 3 - 1 = 1.$

Ex. 1. Evaluate $P(-2)$, $P(0)$, $P(2)$ when $P(x) = 2x^3 + 3x^2 - \frac{5}{2}x + \frac{11}{2}$.

Ex. 2. If $P(x)$ is a quadratic, find $P(x)$ when $P(-2) = 9$, $P(0) = -5$, $P(3) = 7$.

The linear polynomial $ax + b$

Suppose $y = ax + b$, where $a \neq 0$. We have seen that the graph of the locus $\{(x, y) | y = ax + b\}$ is a straight line with gradient a; we say that $y = ax + b$ is the equation of the line. The line meets the x-axis, $y = 0$, at the point where $x = -b/a$, and at no other point.

Suppose now that a certain linear polynomial is zero when $x = p$. What form does the polynomial take? The linear polynomial is

$$P(x) = ax + b.$$

But $P(p) = 0$, so that

$$0 = ap + b$$

or

$$b = -ap.$$

Hence

$$P(x) = ax - ap$$
$$= a(x - p).$$

The value of the constant a cannot be determined without further information. We know the graph of $y = ax + b$ passes through $(p, 0)$; if we know the gradient of the graph, or, what is equivalent, the coordinates of another point on the graph, then a can be found.

For example, suppose we know also that $P(q) = A$. Then

$$A = a(q - p)$$

and so

$$P(x) = \frac{A(x - p)}{q - p}.$$

Ex. 3. Find the linear polynomial which is zero when $x = 2$, and has the value 5 when $x = 4$.

Ex. 4. Find $P(x)$, a linear polynomial, when (i) $P(-2) = 0$, $P(3) = 7$; (ii) $P(\frac{1}{2}) = 0$, $P(-\frac{1}{2}) = -3$.

Identity

Two linear polynomials $ax + b$ and $cx + d$ are identical only when $a = c$ and $b = d$. If we assume that the variable is x, the polynomial $ax + b$ can be regarded as the ordered pair (a, b). There is agreement here with the equality of ordered pairs; $(a, b) = (c, d)$ only if $a = c$ and $b = d$.

In particular $ax + b$ is identically zero only when $a = 0$, $b = 0$.

What can be said about a linear polynomial which has two zeros? Consider, for example,

$$P(x) = \frac{4x-3}{4} - \frac{3x+5}{3} + \frac{29}{12}.$$

We have

$$P(0) = -\frac{3}{4} - \frac{5}{3} + \frac{29}{12} = 0,$$

$$P(1) = \frac{1}{4} - \frac{8}{3} + \frac{29}{12} = 0.$$

Now, since

$$P(1) = 0$$

we have

$$P(x) = a(x-1).$$

But

$$P(0) = 0,$$

so that

$$0 = a(-1) \Leftrightarrow a = 0.$$

Hence

$$P(x) = 0 \quad \text{whatever the value of } x.$$

In general, if $P(p) = P(q) = 0$,

$$P(p) = 0 \Rightarrow \quad P(x) = a(x-p),$$

$$P(q) = 0 \Rightarrow \quad 0 = a(q-p).$$

But, if p and q are distinct, $q - p \neq 0$ and so $a = 0$, and $P(x)$ is identically zero, i.e. $P(p) = P(q) = 0 \Rightarrow P(x) = 0$ for all values of x.

Returning to the particular example,

$$P(x) = \frac{4x-3}{4} - \frac{3x+5}{3} + \frac{29}{12}$$

$$= x - \frac{3}{4} - x - \frac{5}{3} + \frac{29}{12}$$
$$= 0.$$

Clearly $P(x) = 0$ for all values of x.

We therefore see that, if a linear polynomial $P(x)$ is zero for two distinct values, it must be identically zero and so must be zero for every value of x.

Consider the linear polynomial which has the value A when $x = p$ and the value B when $x = q$. This is

$$P(x) = \frac{A(x-q)}{p-q} + \frac{B(x-p)}{q-p}.$$

Furthermore there cannot be two such polynomials, since, if $Q(x)$ were a second linear polynomial satisfying the same conditions,

$$F(x) = P(x) - Q(x)$$

would be a linear polynomial such that $F(p) = 0$, $F(q) = 0$, $p \neq q$, and $F(x)$ would be identically zero.

Note that

$$y = \frac{A(x-q)}{p-q} + \frac{B(x-p)}{q-p}$$

is the equation of the line joining the two points (p, A) and (q, B).

Ex. 5. Show that a linear polynomial $P(x)$ which is such that $P(x_1) = P(x_2) = c$, must be the constant c.

Ex. 6. Write down $P(x)$, a linear polynomial, when (i) $P(2) = 5$, $P(5) = 11$; (ii) $P(-2) = 5$, $P(1) = -3$.

Ex. 7. Write down the equation of the line joining (i) $(3, 5)$, $(-2, 7)$; (ii) $(-3, -2)$, $(-1, 5)$.

Ordered triples in 3-space

A set of three numbers (x, y, z) is an *ordered triple* if

(i) (x, y) is an ordered pair, which we denote by u, and also

(ii) (u, z) is an ordered pair.

An ordered triple can be represented by cartesian coordinates in three dimensions, or 3-space, referred to mutually perpendicular axes Ox, Oy, Oz. The conventional arrangement of the axes is shown in Fig. 7.1, where Ox is meant to be drawn from O,

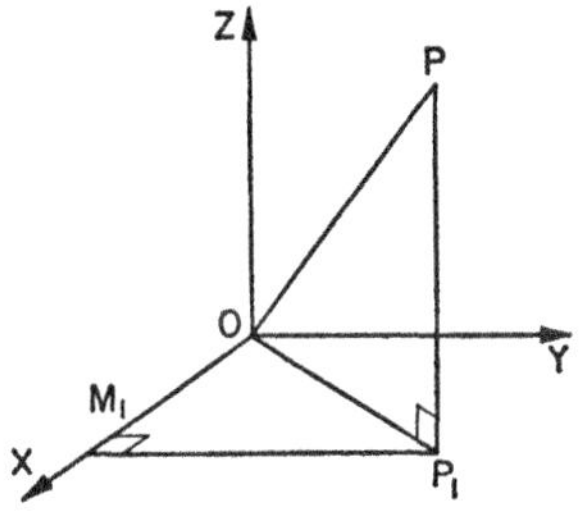

Fig. 7.1

which lies in the plane of the page, outwards towards the reader.

If there is a unique value of z for each ordered pair (x, y), we say that z is a function of x and y, and we write

$$z = f(x, y).$$

The locus of the function is $\{(x, y, z) | z = f(x, y)\}$, and we can often describe the graph of the function, but it is not necessarily easy to construct the graph in three dimensions.

Ex. 8. Describe in words the graphs of the functions (i) $z = x$; (ii) $z = y$; (iii) $z = 1$.

Ex. 9. Show that the distance of (x, y, z) from the origin is $\sqrt{(x^2 + y^2 + z^2)}$. Describe the loci (i) $x^2 + y^2 + z^2 = 1$; (ii) $x^2 + y^2 + z^2 < 1$.

Parametric equations of a line in 3-space

In Fig. 7.2, suppose $P_1(x_1, y_1, z_1)$ and $P_2(x_2, y_2, z_2)$ are two given points, and P is a point on P_1P_2 such that

$$P_1P : P_1P_2 = \lambda : 1,$$

so that

$$P_1P : PP_2 = \lambda : 1 - \lambda.$$

Then

$$\overrightarrow{OP} = \lambda\overrightarrow{OP_2} + (1 - \lambda)\,\overrightarrow{OP_1}.$$

Hence, if P is (x, y, z),

$$x = \lambda x_2 + (1 - \lambda)\,x_1, \qquad y = \lambda y_2 + (1 - \lambda)\,y_1,$$
$$z = \lambda z_2 + (1 - \lambda)\,z_1.$$

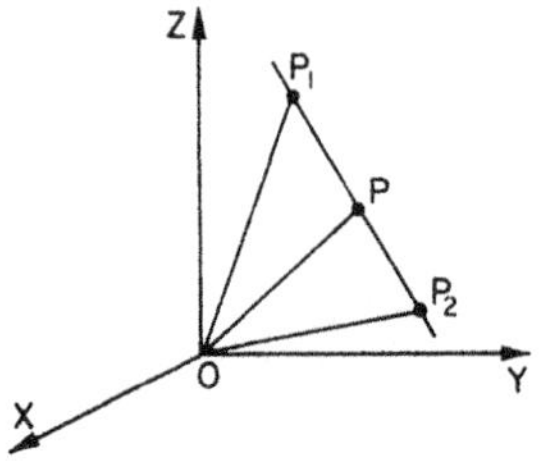

Fig. 7.2

As λ is varied we get different members of the set of points on the line P_1P_2. λ is called the *parameter* of the point P, and the three equations which determine x, y, z in terms of λ, P_1, P_2 are called the parametric equations of the line P_1P_2.

Ex. 10. Write down parametric equations of the lines joining:
(i) (0, 0, 0), (1, 2, 3); (ii) (1, 2, 1), (2, 1, 2); (iii) (−1, 2, 0), (1, −2, 0); (iv) (3, −1, 2), (−1, 3, 5).

Ex. 11. Distinguish between the loci:
(i) $x = 1 - \lambda$, $y = 2\lambda$, $z = 0$; (ii) $x = 1 - \lambda$, $y = 2\lambda$.

EXAMPLE. *Find where the line joining* (2, 1, 0) *to* (−1, 4, 6) *cuts the plane* $4x + y - 3z = 0$.

The parametric equations of the line joining (2, 1, 0) to (−1, 4, 6) are

$$x = -\lambda + (1 - \lambda)\,.\,2, \qquad y = 4\lambda + (1 - \lambda)\,.\,1,$$
$$z = 6\lambda + (1 - \lambda)\,.\,0,$$

which reduce to

$$x = 2 - 3\lambda, \qquad y = 1 + 3\lambda, \qquad z = 6\lambda.$$

This line cuts the plane $4x + y - 3z = 0$ where

$$4(2 - 3\lambda) + (1 + 3\lambda) - 3(6\lambda) = 0,$$

giving

$$\lambda = \tfrac{1}{3}$$

The point of intersection of the line with the plane is therefore

$$(2 - 1, \tfrac{4}{3} + \tfrac{2}{3}, 2), \quad \text{or} \quad (1, 2, 2).$$

Linear polynomial in two variables

The linear polynomial in two independent variables x and y is $P(x, y) = ax + by + c$, where a, b, c are coefficients.

We shall now consider the locus $\{(x, y, z) | z = ax + by + c\}$ and its graph, whose equation is $z = ax + by + c$.

Suppose $P_1(x_1, y_1, z_1)$ and $P_2(x_2, y_2, z_2)$ are any two points on the graph of $z = ax + by + c$, so that

$$z_1 = ax_1 + by_1 + c, \qquad z_2 = ax_2 + by_2 + c.$$

Any point $P(x, y, z)$ on the straight line P_1P_2 has the parametric equations.

$$x = \lambda x_2 + (1 - \lambda) x_1, \qquad y = \lambda y_2 + (1 - \lambda) y_1,$$
$$z = \lambda z_2 + (1 - \lambda) z_1.$$

Hence $z - (ax + by + c)$

$$= \lambda z_2 + (1 - \lambda) z_1$$
$$- a\{\lambda x_2 + (1 - \lambda) x_1\} - b\{\lambda y_2 + (1 - \lambda) y_1\} - c$$
$$= \lambda(z_2 - ax_2 - by_2 - c) + (1 - \lambda)(z_1 - ax_1 - by_1 - c)$$
$$= 0.$$

Therefore $z = ax + by + c$, so that, if P_1 and P_2 are two points on the graph, any point on the straight line P_1P_2 also lies on the graph. This is the characteristic property of a plane surface, and so $z = ax + by + c$ is the equation of a plane.

Ex. 12. Verify that (1, 1, 2) and (3, 2, 1) lie on the graph of $z = 3x - 7y + 6$. Verify that every point on the straight line joining these two points also lies on $z = 3x - 7y + 6$. What conclusion do you draw?

The sign of $ax + by + c$

In Fig. 7.3, suppose AB is the intersection of the plane $z = ax + by + c$ with the plane $z = 0$. The equation of AB, in the plane XOY, is

$$ax + by + c = 0.$$

AB divides the plane XOY into two areas.

The plane $z = ax + by + c$ lies above all points in one of

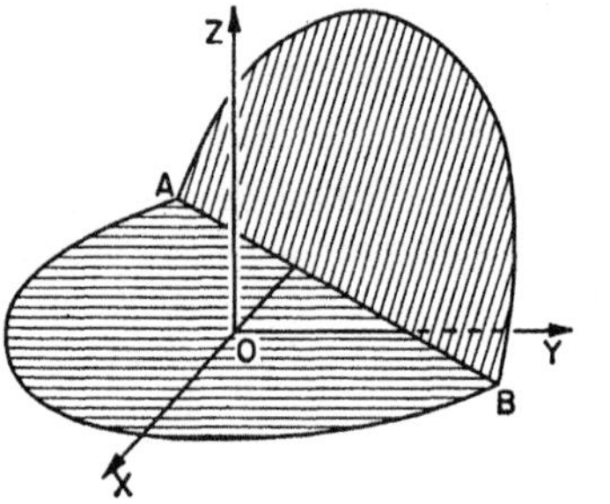

Fig. 7.3

these areas, and below all points in the other area. In other words, for all points (x, y) on one side of AB the expression $ax + by + c$ is positive, and for all points (x, y) on the other side of AB the expression $ax + by + c$ is negative.

Consider, for example, Fig. 7.4, where AB is the line

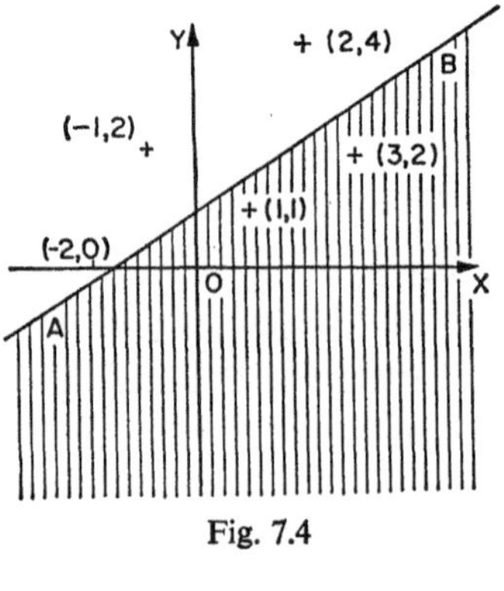

Fig. 7.4

$3x - 4y + 5 = 0$. If $P(x, y) = 3x - 4y + 5$ then

$$P(0, 0) = 5, \quad P(1, 1) = 4, \quad P(3, 2) = 6,$$
$$P(-2, 0) = -1, \quad P(-1, 2) = -6, \quad P(2, 4) = -5.$$

In Fig. 7.5, the line AB represents the graph of $ax + by + c = 0$. AB divides the plane into two areas which are called *open half*

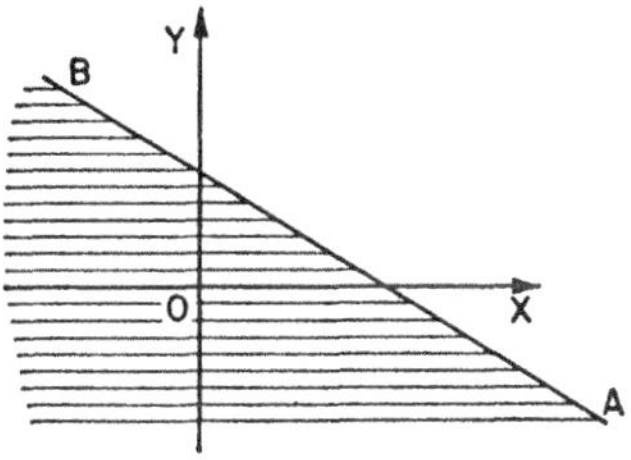

Fig. 7.5

planes. In one half plane $ax + by + c < 0$, and in the other half plane $ax + by + c > 0$. The union of a half plane and its boundary line is called a *closed half plane*.

For example, suppose that, in Fig. 7.5, the line AB is $2x + 3y - 5 = 0$. Since the value of $2x + 3y - 5$ at O is (-5), the shaded open half plane is the locus $\{(x, y)|2x + 3y - 5 < 0\}$, and the unshaded open half plane is the locus

$$\{(x, y)|2x + 3y - 5 > 0\}.$$

The locus $\{(x, y)|2x + 3y - 5 \leqslant 0\}$ is the closed half plane that is shaded, that is, the shaded area together with the line AB.

Ex. 13. Sketch figures and shade them to represent the loci:
(i) $x - y > 0$; (ii) $x + y < 0$; (iii) $2x + 3y - 5 < 0, x < 0$;
(iv) $3x - 2y + 4 < 0, y > 0$;
(v) $3x + y - 3 < 0, x + 3y - 6 < 0$;
(vi) $x > 0, y < 0, 2x - 3y + 6 > 0$;
(vii) $4x + 3y - 12 > 0, x + 2y - 4 < 0, x - 2y - 4 < 0$.

Linear programming

Consider the following problem:

A poultry farmer needs to provide 30 gm of chemical A and 40 gm of chemical B every day in order to keep his hens healthy.

It is both difficult and expensive to buy the pure chemicals, but he can buy a food F_1 which has 0·06 per cent of A and 0·04 per cent of B and another food F_2 which has 0·05 per cent of A and 0·08 per cent of B. The prices of F_1 and F_2 are 12*s.* per kg and 15*s.* per kg respectively. How can he both keep his hens healthy and spend as little as possible on the special foods?

Suppose he buys x kg of F_1 and y kg of F_2 each day, and let C shillings be the cost of these foods. Then

$$C = 12x + 15y.$$

The problem, however, is not to solve an equation but to make C as small as possible (to minimize C). The conditions are:

$$60x + 50y \geqslant 30, \qquad \text{(i)}$$

$$40x + 80y \geqslant 40. \qquad \text{(ii)}$$

The locus (i) is the shaded closed half plane in Fig. 7.6. The locus (ii) is the shaded closed half plane in Fig. 7.7.

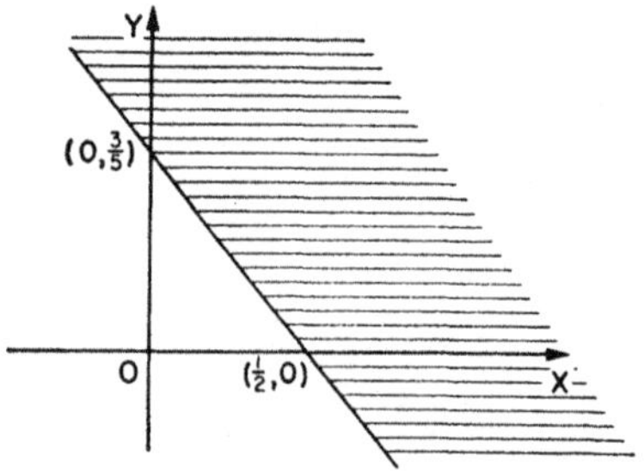

Fig. 7.6

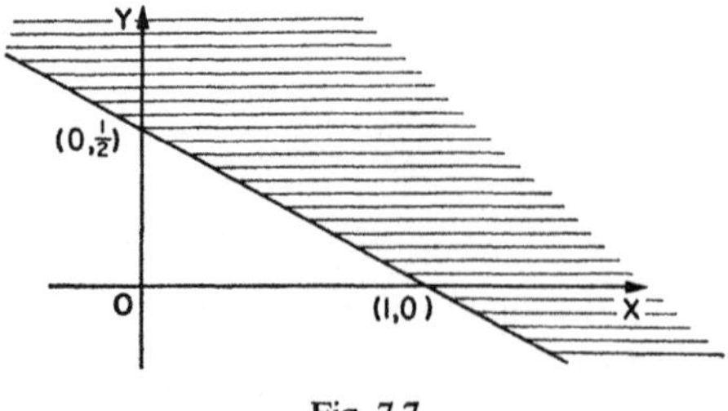

Fig. 7.7

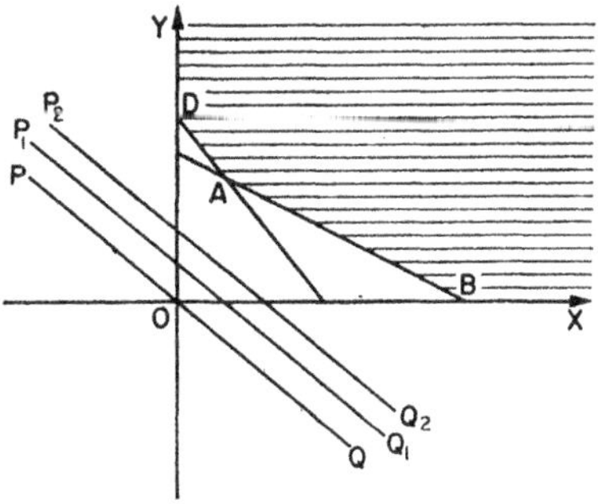

Fig. 7.8

We must also have $x > 0$, $y > 0$, and the intersection of the four loci concerned with the conditions of the problem is shown in Fig. 7.8. In this figure the line PQ represents $C = 0$, or $12x + 15y = 0$. If we move PQ parallel to itself towards A, into positions like P_1Q_1 or P_2Q_2, we get lines along which C is constant, but the value of C for each line is steadily increasing. But we do not satisfy the conditions of the problem unless (x, y) lies within the shaded area or along AB or along AD.

Clearly the least value of C consistent with the conditions is obtained by drawing the line $12x + 15y = C$ through A.

A satisfies $6x + 5y = 3$, $x + 2y = 1$, and so is $(\frac{1}{7}, \frac{3}{7})$, and so the least cost of the foods is

$$12 \times \frac{1}{7} + 15 \times \frac{3}{7} = \frac{57}{7} \approx 8s.\ 2d. \text{ per day,}$$

or £2 17*s*. per week.

Ex. 14. Find the least value of $x + y$ when $x > 0$, $y > 0$, $5x + 4y \geqslant 20$, $3x + 7y \geqslant 21$.

Ex. 15. Find the least value of $5x + 3y$ when $x > 0$, $y > 0$, $4x + y \geqslant 8$, $5x + 6y \geqslant 15$.

Ex. 16. Find the greatest value of $3x + 5y$ when $x > 0$, $y > 0$, $3x + 2y \leqslant 12$, $x + 2y \leqslant 8$.

Ex. 17. Find the greatest value of $2x + 3y$ when $x > 0$, $y > 0$, $x + 3y \leqslant 15$, $x + y \leqslant 7$, $2x + y \leqslant 6$.

EXERCISE 7.1

1. Sketch the line $x - 2y + 4 = 0$ and plot the points $(-4, -1)$, $(-1, 1)$, $(3, 3)$, $(5, 4)$, $(-4, 1)$, $(-1, 2)$, $(3, 4)$, $(5, 5)$.
Verify from the sketch that the first four points lie on the same side of the line as the origin, and the last four on the opposite side. Evaluate the function $(x - 2y + 4)$ for each of the eight points.

2. Find the locus of points which are equidistant from the lines $y = 2$ and $4x + 3y = 10$.

3. Find the coordinates of the centre and the radius of the circle inscribed in the triangle formed by the lines $y + 2 = 0$, $3x - 4y + 28 = 0$, $4x + 3y = 46$.

4. Describe the locus $z \geqslant 0$, $x^2 + y^2 + z^2 \leqslant 1$.

5. Sketch the locus $x > 0$, $y > 0$, $x + 3y \leqslant 15$, $x + y \leqslant 7$, $2x + y \leqslant 12$.

6. Sketch the locus $3x - 5y \leqslant 0$, $2x - y \geqslant 0$, $x + 3y \leqslant 14$, and calculate the area enclosed.

7. I can buy coffee at \$2.40 per kg and tea at \$2 per kg. From 1 kg of coffee I make 30 l. of coffee, and from 1 kg of tea I make 45 l. I drink more than twice as much coffee as tea. If I do not want to spend more than \$13.80 what is the greatest weight of tea I can buy?

8. In buying chocolates I decide to buy either x kg at 4 fr per kg and y kg at 5 fr per kg or x kg at 8 fr per kg and y kg at 3 fr per kg. I do not want to spend more than 42 fr. What is the greatest weight of chocolates I can buy?

9. A benefactor leaves \$1,000 to divide between 50 children of ages 3, 4 and 5. He says that no child is to have less than \$16, that the amount per child is to be proportional to the number of years in his or her age, that there are not to be more 5-year-olds than 4-year-olds, and not more 4-year-olds than 3-year-olds. Find how the money will be distributed if as many as possible of the children are to be (i) 3 years old, (ii) 4 years old, (iii) 5 years old.

10. A tea blender has 2,100 kg of blend A, 1,900 kg of B and 1,200 kg of C. He mixes the blends so that each case of type X contains 2 kg of A, 3 kg of B and 2 kg of C, and each case of type Y contains 3 kg of A, 2 kg of B and 1 kg of C. He makes a profit of 6 fr on each case of X and 8 fr on each case of Y. Find his maximum profit.

11. If A is the point $(0, 0, 6)$ and P is any point $(a, b, 0)$ in the plane $z = 0$, write down the point K such that $AK{:}AP = \lambda{:}1$. If λ is constant and a, b are variable write down the locus of K.

12. Find the equations of the planes through (i) $(3, 0, 0)$, $(0, 4, 0)$, $(1, 3, 1)$ and (ii) $(0, 0, 6)$, $(2, 0, 2)$, $(1, 2, 1)$. Find in its simplest form the equation of the line of intersection of these planes.

13. If $P(x, y, z) = 3x - 4y + 12z - 12$, find the values of $P(4, 0, 0)$, $P(0, 0, 1)$, $P(0, 0, 0)$, $P(4, -3, 1)$. Sketch the plane $3x - 4y + 12z = 12$ and indicate the graph of the locus $3x - 4y + 12z \leqslant 12$.

14. Find an expression for the distance of (x, y, z) from the line $x - 2y = 0$, $z = 0$. Hence state what is represented by the locus $(x - 2y)^2 + 5z^2 \leqslant 5$.

CHAPTER 8

THE QUADRATIC FUNCTION

Graph of a quadratic function

Figure 8.1 shows the graph of

$$y = x^2 - 2x - 3.$$

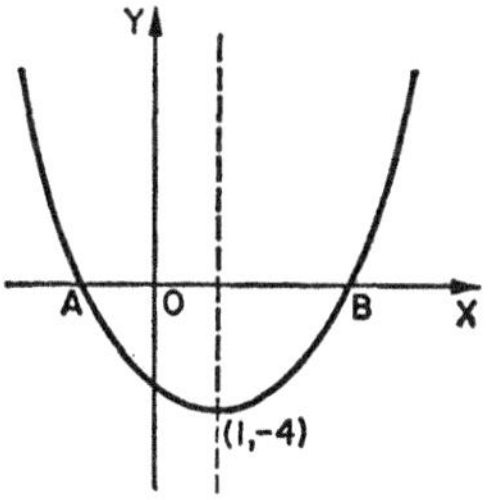

Fig. 8.1

Since $x^2 - 2x - 3 = (x + 1)(x - 3)$, the graph cuts Ox at $A(-1, 0)$ and $B(3, 0)$. The curve is symmetrical about $x = 1$ and its lowest point is $(1, -4)$.

Figure 8.2 shows the graph of

$$y = 4x^2 - 12x + 9.$$

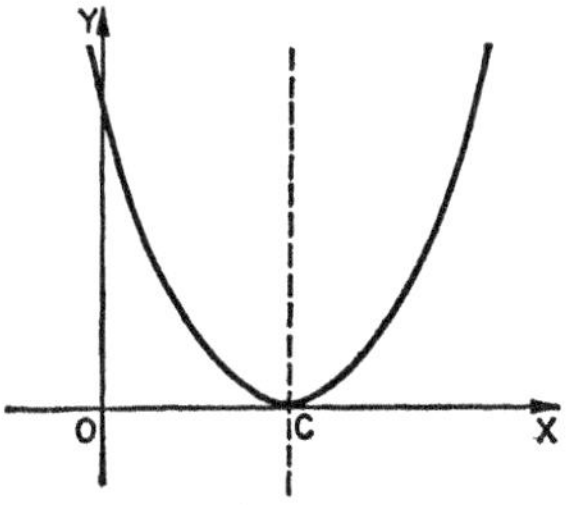

Fig. 8.2

Since $4x^2 - 12x - 9 = (2x - 3)^2$, the graph touches Ox at C (1·5, 0). The curve is symmetrical about $x = 1{\cdot}5$ and C is its lowest point. The function $(2x - 3)^2$ can be zero but cannot be negative.

Figure 8.3 shows the graph of

$$y = 4x - 5 - x^2.$$

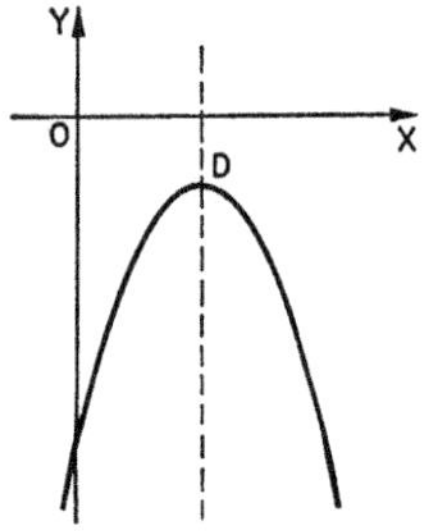

Fig. 8.3

Since $4x - 5 - x^2 = -1 - (x - 2)^2$, the value of y cannot be greater than -1 and the graph does not cut Ox. The curve is symmetrical about $x = 2$ and D (2, -1) is its highest point.

Ex. 1. Sketch the graphs of (i) $y = 8 - 2x - x^2$, (ii) $y = 4x - 4x^2 - 1$, (iii) $y = 2x^2 - 3x + 2$. In each case state the axis of symmetry and a highest or lowest point.

The sign of $(x - a)(x - b)$

Suppose $a < b$ and let A, B represent a, b as in Fig. 8.4. When

$$x < a, \quad (x - a) < 0,\ (x - b) < 0 \Rightarrow (x - a)(x - b) > 0,$$
$$a < x < b, \quad (x - a) > 0,\ (x - b) < 0 \Rightarrow (x - a)(x - b) < 0,$$
$$b < x, \quad (x - a) > 0,\ (x - b) > 0 \Rightarrow (x - a)(x - b) > 0.$$

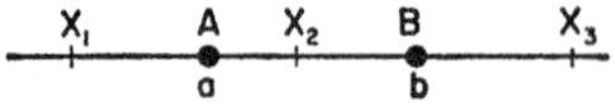

Fig. 8.4

Using set notation, the solution set of the inequation $(x - a)(x - b) > 0$ is the union

$$\{x \mid x < a\} \cup \{x \mid x > b\}.$$

The solution set of the inequation $(x - a)(x - b) < 0$ is

$$\{x \mid a < x < b\}.$$

Graphical solution of inequalities

Consider the set of all real points (x, y) in a plane and the subsets:

$$R\ \{(x, y) | y < 2 + 3x - 2x^2\},$$
$$T\ \{(x, y) | y = 0\}.$$

In Fig. 8.5, R is represented by the shaded area, and T by the line XOX'. Now $R \cap T$ gives $0 < 2 + 3x - 2x^2$, and in the figure is represented by the line-segment AB. Thus

$$2 + 3x - 2x^2 > 0 \text{ when } -\tfrac{1}{2} < x < 2.$$

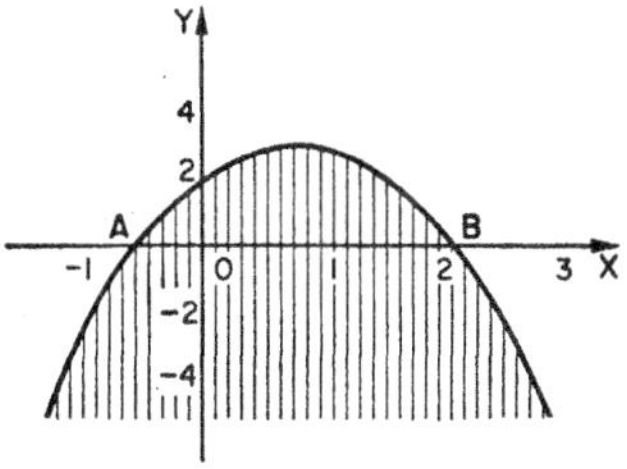

Fig. 8.5

Ex. 2. By sketching graphs solve the inequations (i) $2x^2 + x - 6 > 0$, (ii) $10 + 11x - 6x^2 > 0$.

EXERCISE 8.1

In nos. 1–6, write down the range of values of x for which the given function is (i) positive, (ii) negative, using a graphical method if it helps.

1. $(x - 1)(x - 2)$. **2.** $x(x + 2)$. **3.** $(x + 2)(1 - x)$.

4. $2x^2 + x - 3$. **5.** $6 + x - 2x^2$. **6.** $4x^2 - 12x + 9$.

In nos. 7–12 *the given function of x is of constant sign. Find whether the function is positive or negative, its greatest or least value and the corresponding value of x.*

7. $x^2 + 2x + 3$. **8.** $4x - 5 - x^2$. **9.** $4x^2 + 4x + 5$.

10. $x - 1 - x^2$. **11.** $-20x - 4x^2 - 28$. **12.** $\frac{1}{2} - 2x + 4x^2$.

Describe the nature of the function in nos. 13–18.

13. $2x^2 + 5x + 2$. **14.** $2x^2 + 4x + 2$. **15.** $2x^2 + 3x + 2$.

16. $-4x - 6 - 4x^2$. **17.** $5 + 9x - 2x^2$. **18.** $30x - 25 - 9x^2$.

In nos. 19–24 *write down the solution sets of the inequations.*

19. $(x - 2)(x - 5) > 0$. **20.** $(x + 1)(x - 3) < 0$.

21. $(x + 3)(x - 2) > 0$. **22.** $2 - x - x^2 > 0$.

23. $6 + 5x + x^2 > 0$. **24.** $9 - 4x^2 > 0$.

The function $y = ax^2 + bx + c$

$$ax^2 + bx + c = a\left(x^2 + \frac{b}{a}x + \frac{c}{a}\right)$$

$$= a\left\{\left(x + \frac{b}{2a}\right)^2 + \frac{4ac - b^2}{4a^2}\right\}$$

(I) $a > 0$

(i) $b^2 < 4ac$. Then

$$\left(x + \frac{b}{2a}\right)^2 \geqslant 0 \quad \text{and} \quad \frac{4ac - b^2}{4a^2} > 0,$$

and so $ax^2 + bx + c > 0$; the graph of y is like Fig. 8.6 (i).

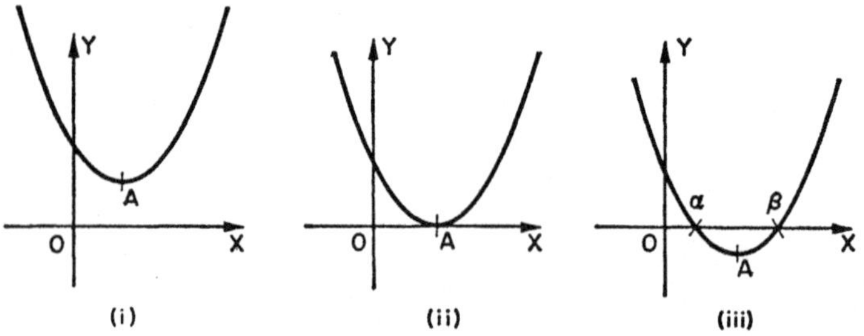

Fig. 8.6

(ii) $b^2 = 4ac$. Then

$$y = a\left(x + \frac{b}{2a}\right)^2 \geqslant 0;$$

the graph is like Fig. 8.6 (ii).

(iii) $b^2 > 4ac$. Let $(b^2 - 4ac)/4a^2 = p^2$; then

$$y = a\left(x + \frac{b}{2a} + p\right)\left(x + \frac{b}{2a} - p\right).$$

Thus y is zero when

$$x = -\frac{b}{2a} - p \quad \text{or} \quad x = -\frac{b}{2a} + p.$$

Denoting these two roots by α, β $(\alpha < \beta)$

$$y = a(x - \alpha)(x - \beta),$$

and the graph is like Fig. 8.6 (iii). Note that in (iii),

$$ax^2 + bx + c > 0$$

except when $\alpha < x < \beta$, and then $ax^2 + bx + c < 0$.

In all cases the curve is symmetrical about $x = -b/2a$ and the lowest point is $A\ \{-b/2a, (4ac - b^2)/4a\}$.

(II) $a < 0$

As before $y = ax^2 + bx + c = a\left\{\left(x + \frac{b}{2a}\right)^2 + \frac{4ac - b^2}{4a^2}\right\}$

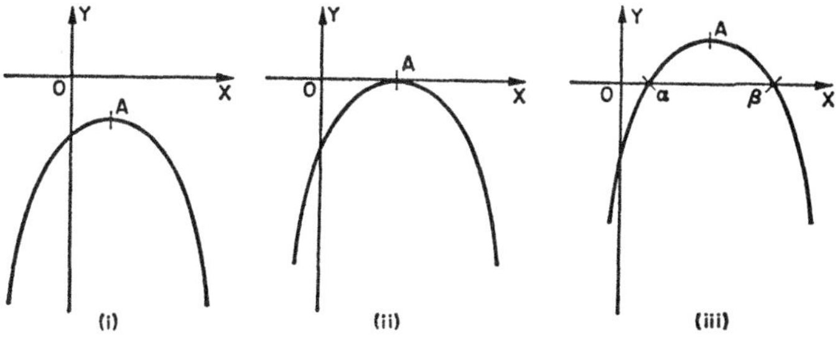

Fig. 8.7

The only change necessary when $a < 0$ is a change in the sign of y. The results are shown in Fig. 8.7 (i), (ii), (iii).

Note that, when $a < 0$ and $b^2 > 4ac$, from Fig. 8.7 (iii),

$$ax^2 + bx + c > 0 \quad \text{only when} \quad \alpha < x < \beta.$$

In all cases the maximum value of y occurs when $x = -b/2a$.

EXAMPLE. *Find any limitations which may exist in the value of the function*

$$y = \frac{x^2 - 5}{x^2 - x - 2}.$$

For a given value of y the corresponding values of x are given by

$$y(x^2 - x - 2) = x^2 - 5,$$

$$(y - 1)\,x^2 - yx - (2y - 5) = 0.$$

The values of x are real only when $(b^2 \geqslant 4ac)$

$$y^2 \geqslant -4(y - 1)(2y - 5)$$

$$= -4(2y^2 - 7y + 5),$$

$$9y^2 - 28y + 20 \geqslant 0,$$

$$9\,(y - \tfrac{10}{9})(y - 2) \geqslant 0.$$

The value, y, of the given function, therefore cannot lie between $\frac{10}{9}$ and 2.

A sketch of the graph of the function is shown in Fig. 8.8.

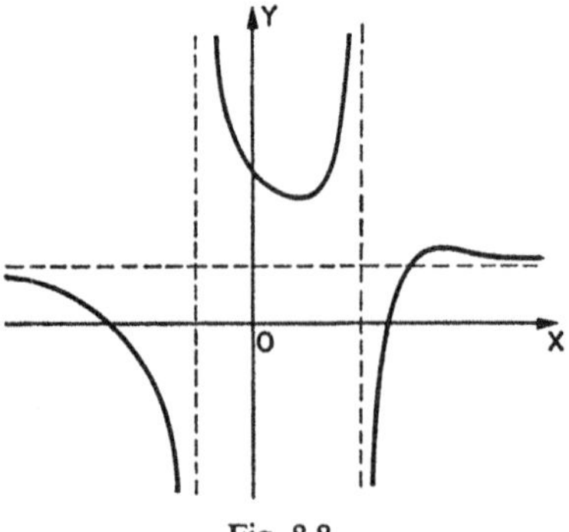

Fig. 8.8

EXERCISE 8.2

1. Sketch the graphs of $y = x^2$ and $y = x$. Shade the area between the two curves where $x^2 < x$.

2. Sketch and colour the set A of points (x, y) such that $y < 6 - x - x^2$, and the set B of points (x, y) where $y = -1$. What is $A \cap B$ and when is $-1 < 6 - x - x^2$?

When is (i) $x^2 + x < 6$, (ii) $x^2 + x < 2$, (iii) $x^2 + x < 6{\cdot}25$?

3. Show that $3x^2 + 9x + 7$ is always positive. What value of x makes the value of the function least, and what is this least value?

4. Show that $4x^2 - 8x + 7$ is always positive. Find its minimum value and the corresponding value of x.

5. Prove that $4x - 2x^2 - 3$ is always negative. Find its greatest value and the corresponding value of x.

6. Show that $(x - 1)(x - 2)/(x - 3)$ can take all values except those lying between $3 - 2\sqrt{2}$ and $3 + 2\sqrt{2}$.

7. Find the values of x for which $6 - 7x - 3x^2$ is negative, and find its greatest value.

8. Prove that the function $x^2/(x^2 - 2x + 2)$ always lies between 0 and 2. Sketch the general shape of its graph.

9. Find if there are any limitations to the value of $(x^2 - 1)/(x + 2)$, and sketch the graph of the function.

10. Find the range of values of k if the function $8x^2 + 4kx + 3k$ is positive for all values of x.

11. Find the quadratic function of x which is positive only when $1 < x < 3$ and which takes the value 4 when $x = 2$.

12. Find the quadratic function of x which is negative only when $-2 < x < 1$ and which takes the value 12 when $x = 2$.

13. Find the range of values of a necessary to ensure that the function $1 + ax + a(a + 1)x^2$ is positive for all real values of x.

14. Two real variables x and y are connected by the equation

$$(2x + y)^2 - 4(x + y) + 1 = 0.$$

Prove that (i) $y \geqslant 0$, (ii) $x \leqslant \frac{3}{4}$.

15. Find the range of values of x for which $5x^2 - 14x + 9$ is negative.

Find the values of k for which the expression $5x^2 - 14x + 9 + k(x^2 - 2x)$ is a perfect square.

Show that, if k lies between these two values, the expression is never zero.

The quadratic equation—formula

The formula for the solution of a quadratic equation is established as follows:

The standard equation is

$$ax^2 + bx + c = 0.$$

Then

$$x^2 + \frac{b}{a}x = -\frac{c}{a}.$$

Therefore

$$x^2 + \frac{b}{a}x + \left(\frac{b}{2a}\right)^2 = \left(\frac{b}{2a}\right)^2 - \frac{c}{a},$$

$$\left(x + \frac{b}{2a}\right)^2 = \frac{b^2 - 4ac}{4a^2},$$

$$x + \frac{b}{2a} = \frac{\pm\sqrt{(b^2 - 4ac)}}{2a},$$

$$x = \frac{-b \pm \sqrt{(b^2 - 4ac)}}{2a}.$$

If α, β are the roots of $ax^2 + bx + c = 0$ we can write

$$\alpha = \frac{-b}{2a} + \frac{\sqrt{(b^2 - 4ac)}}{2a}, \quad \beta = \frac{-b}{2a} - \frac{\sqrt{(b^2 - 4ac)}}{2a}.$$

If $b^2 > 4ac$, both α and β exist and are unequal.

If $b^2 = 4ac$, both α and β exist and are equal.

If $b^2 < 4ac$, neither α nor β can be evaluated because $\sqrt{(b^2 - 4ac)}$ cannot be found. There is, however, a form of algebra called *complex algebra* in which the expressions for α and β have an interpretation.

Compare these results with Figs. 8.6 and 8.7, on pp. 112, 113.

A theorem concerning identities

Suppose y is a linear function of x, say,

$$y = Ax + B.$$

If y is zero when $x = p$ and when $x = q$, and $p \neq q$, then $A = 0$, $B = 0$ and y is zero for all values of x.

We have

$$Ap + B = 0, \quad Aq + B = 0$$

and so

$$A(p - q) = 0.$$

But $p - q \neq 0$, thus $A = 0$, and so also $B = 0$. Therefore y must be zero for all values of x.

Roots and coefficients of a quadratic equation

Suppose that the roots of $ax^2 + bx + c = 0$ are α, β. Consider the function y of x where

$$y = ax^2 + bx + c - a(x - \alpha)(x - \beta).$$

Then y is a linear function of x, namely

$$y = x\{b + a(\alpha + \beta)\} + \{c - a\alpha\beta\}.$$

But $ax^2 + bx + c$ is zero when $x = \alpha$ and when $x = \beta$. Hence y is zero when $x = \alpha$ and when $x = \beta$, and so

$$b + a(\alpha + \beta) = 0, \quad c - a\alpha\beta = 0.$$

Thus

$$\alpha + \beta = -\frac{b}{a}, \quad \alpha\beta = \frac{c}{a}.$$

Since y is always zero, it also follows that

$$ax^2 + bx + c \equiv a(x - \alpha)(x - \beta).$$

When

$$\alpha = \beta, \quad b^2 = 4ac, \text{ and } \alpha = \beta = -b/2a.$$

The relations $\alpha + \beta = -b/a$, $\alpha\beta = c/a$ are therefore still true when $\alpha = \beta$.

Conversely, the quadratic equation with roots α, β must be

$$(x - \alpha)(x - \beta) = 0, \text{ or } x^2 - (\alpha + \beta)\,x + \alpha\beta = 0.$$

For, if $ax^2 + bx + c = 0$ is another quadratic equation with roots a, β then

$$x^2 + \frac{b}{a} x + \frac{c}{a} - (x - \alpha)(x - \beta) = 0$$

is a linear equation with two distinct roots, and so is identically zero. Thus $x^2 + (b/a) x + (c/a)$ is the same function as

$$(x - \alpha)(x - \beta).$$

EXAMPLE. *If α and β are the roots of the equation*

$$3x^2 + 5x - 4 = 0,$$

find the value of (i) $\alpha^2 + \beta^2$, (ii) $\alpha^3 + \beta^3$.

Since α, β are the roots of $3x^2 + 5x - 4 = 0$ we have

$$\alpha + \beta = -\tfrac{5}{3}, \quad \alpha\beta = -\tfrac{4}{3}.$$

(i) But $(\alpha + \beta)^2 = \alpha^2 + 2\alpha\beta + \beta^2$, therefore

$$\alpha^2 + \beta^2 = (\alpha + \beta)^2 - 2\alpha\beta$$

$$= \frac{25}{9} + \frac{8}{3}$$

$$= \frac{49}{9}.$$

(ii) Also $(\alpha + \beta)^3 = \alpha^3 + 3\alpha^2\beta + 3\alpha\beta^2 + \beta^3$, therefore

$$\alpha^3 + \beta^3 = (\alpha + \beta)^3 - 3\alpha\beta(\alpha + \beta)$$

$$= -\frac{125}{27} + 4\left(-\frac{5}{3}\right)$$

$$= \frac{-125 - 180}{27}$$

$$= -\frac{305}{27}.$$

Ex. 3. If α, β ($\alpha < \beta$) are the roots of $x^2 - 4x + 2 = 0$, find the value of (i) $(\alpha - \beta)^2$; (ii) $\alpha^2 + \alpha\beta + \beta^2$; (iii) $\alpha^3 - \beta^3$; (iv) $\alpha^4 + \beta^4$.

EXAMPLE. *If* α *and* β *are the roots of the equation*

$$2x^2 - 3x - 7 = 0,$$

form the equation whose roots are (*i*) $\alpha + 1$ *and* $\beta + 1$, (*ii*) a^2 *and* β^2.

(i) Suppose $p = \alpha + 1$, $q = \beta + 1$. The equation with roots p and q is

$$(x - p)(x - q) = 0$$

or

$$x^2 - (p + q)\,x + pq = 0.$$

Now

$$\begin{aligned} p + q &= (\alpha + 1) + (\beta + 1) \\ &= (\alpha + \beta) + 2 \\ &= \frac{3}{2} + 2 = \frac{7}{2}; \end{aligned}$$

$$\begin{aligned} pq &= (\alpha + 1)(\beta + 1) \\ &= \alpha\beta + (\alpha + \beta) + 1 \\ &= -\frac{7}{2} + \frac{3}{2} + 1 = -1. \end{aligned}$$

So the required equation is

$$x^2 - \tfrac{7}{2}x - 1 = 0$$

or

$$2x^2 - 7x - 2 = 0.$$

(ii) Suppose $p = \alpha^2$, $q = \beta^2$. The equation with roots p and q is

$$x^2 - (p + q)\,x + pq = 0.$$

Now

$$\begin{aligned} p + q = \alpha^2 + \beta^2 &= (\alpha + \beta)^2 - 2\alpha\beta \\ &= \frac{9}{4} + 7 = \frac{37}{4}, \end{aligned}$$

and

$$pq = (\alpha\beta)^2 = \frac{49}{4}.$$

Therefore the required equation is

$$x^2 - \frac{37}{4}x + \frac{49}{4} = 0$$

or

$$4x^2 - 37x + 49 = 0.$$

The method used in the above solutions is based on the calculation of the sum and product of the required roots. Another method based on a substitution is now shown.

(i) The required equation is

$$\{y - (\alpha + 1)\}\,\{y - (\beta + 1)\} = 0$$

or

$$\{(y - 1) - \alpha\}\,\{(y - 1) - \beta\} = 0.$$

But the given equation was $(x - \alpha)(x - \beta) = 0$, and so we find the required equation by putting $(y - 1)$ instead of x. The result is

$$2(y - 1)^2 - 3(y - 1) - 7 = 0,$$

$$2y^2 - 7y - 2 = 0, \text{ as before.}$$

(ii) The required equation is $(y - \alpha^2)(y - \beta^2) = 0$. Now $2x^2 - 3x - 7 = 0$ is satisfied by $x = \alpha$, $x = \beta$. It can be written $2x^2 - 7 = 3x$, so that α, β are also roots of $(2x^2 - 7)^2 = 9x^2$.

Hence α^2, β^2 are roots of $(2y - 7)^2 = 9y$, which must be the required equation. We can therefore form the required equation by putting $x^2 = y$; to do this we first write the original equation in the form $(2x^2 - 7)^2 = 9x^2$.

Alternatively we can regard the problem as the elimination of x from the simultaneous equations $2x^2 - 3x - 7 = 0$ and $x^2 = y$. This aspect is useful in more complicated problems of the transformation of the roots of a given equation.

Ex. 4. If α, β are the roots of the equation $2x^2 + 5x + 1 = 0$, find the equations with roots (i) $\alpha - 2, \beta - 2$; (ii) $2\alpha^2, 2\beta^2$. Check by using two different methods.

EXAMPLE. *If, in the equation $x^2 + cx + d = 0$, the roots differ by 3, prove that $c^2 = 4d + 9$.*

Suppose the roots are α, β and that

$$\alpha - \beta = 3.$$

Thus

$$(\alpha + \beta)^2 = (\alpha - \beta)^2 + 4\alpha\beta$$
$$= 9 + 4\alpha\beta.$$

But

$$\alpha + \beta = -c, \quad \alpha\beta = d,$$

so

$$c^2 = 9 + 4d.$$

EXERCISE 8.3

1. The roots of $4x^2 - 2x - 7 = 0$ are α and β. Without solving the equation, find the values of $\alpha^2 + \beta^2$ and $\alpha^3 + \beta^3$.

2. The difference between the roots of the quadratic equation $x^2 - 2px + c = 0$ is 2. Express c in terms of p.

3. If α and β are the roots of the equation

$$ax^2 + bx + c = 0,$$

express, in terms of α and β, the roots of the equations:
(i) $ax^2 - bx + c = 0$; (ii) $cx^2 + bx + a = 0$.
Form the equation whose roots are $\alpha + \beta$ and $\alpha\beta$.

4. If the quadratic equations

$$3x^2 + px + 1 = 0 \text{ and } 2x^2 + qx + 1 = 0$$

have a common root, show that

$$2p^2 + 3q^2 - 5pq + 1 = 0.$$

5. The roots of the equation $x^2 - 3x + k = 0$ are α and β. Find the equation whose roots are $\alpha + 1/\beta$ and $\beta + 1/\alpha$, expressing its coefficients in terms of k.
Also express $(\alpha + 1/\beta)^2 + (\beta + 1/\alpha)^2$ in terms of k, and find the possible values of k when $\alpha + 1/\beta = \beta + 1/\alpha$.

6. Show that, if one root of the equation $x^2 + px + q = 0$ is double the other, then $2p^2 = 9q$.

7. If α, β are the roots of the equation $x^2 - 5x + 2 = 0$, write down the values of $\alpha + \beta$ and $\alpha\beta$. Deduce the value of $2(\alpha/\beta + \beta/\alpha)$, and write down the equation whose roots are $2\alpha/\beta$ and $2\beta/\alpha$.

8. For what range of values of a are the roots of the equation $2x^2 - 9x + a = 0$ real?

9. If α, β are the roots of the equation $2x^2 - 6x - 3 = 0$, write down the values of $\alpha + \beta$ and $\alpha\beta$. Deduce the value of $1/\alpha^2 + 1/\beta^2$, and find the equation whose roots are $1/\alpha^2$, $1/\beta^2$.

10. Show that the quadratic equation

$$Kx^2 - 12x + K - 5 = 0$$

has equal roots for two different values of K and find these values.

11. If α and β are the roots of $2x^2 + 5x - 2 = 0$, find the equation whose roots are α/β and β/α.

12. If α and β are the roots of the equation

$$x^2 + bx + c = 0,$$

find the equation whose roots are $\alpha^2 + 1$ and $\beta^2 + 1$.

13. Verify that $x = a$ is one root of the equation $(x - b)^2 - (x - a)(a + b) - (a - b)^2 = 0$ and find the other root.

14. Show that the roots of $x^2 - (p + 2)x + p = 0$ are real for all real values of p.

If α and β are the roots of this equation, form the quadratic equation whose roots are $(\alpha - \beta)^2$ and $4\alpha\beta$.

Hence, or otherwise, show that $(p + 2)^4 - 16p(p^2 + 4)$ is positive or zero for all real values of p.

15. Prove that, if a is real, the expression $a^2 - 4a + 8$ is always positive, and show that the equation $x^2 - ax + a = 2$ has two real distinct roots.

16. Show that the sum of the squares of the roots of the equation $2x^2 + 2px + (p^2 - 8) = 0$ is equal to 8 whatever the value of p. If the value of p is a certain negative whole number the sum of the cubes of the roots of the equation is 20. Determine this value for p.

For what range of values of p are the roots of the equation $2x^2 + 2px + (p^2 - 8) = 0$ real?

17. If a, b, c are real, prove that α and β, the roots of the equation $4(a - y)(c - y) = b^2$ are also real.

Prove that the equation $ax^2 + bx + c = y(x^2 + 1)$ has real roots only when y lies between α and β.

Complex numbers

Consider the quadratic equations

(i) $x^2 - 4x + 3 = 0$, (ii) $x^2 - 4x - 5 = 0$,

(iii) $x^2 - 4x - \frac{9}{4} = 0$, (iv) $x^2 - 4x + 2 = 0$,

(v) $x^2 - 4x + 5 = 0$.

The roots of equations (i) to (iv) are $\{1, 3\}$, $\{-1, 5\}$, $\{-\frac{1}{2}, \frac{9}{2}\}$ and $\{2 - \sqrt{2}, 2 + \sqrt{2}\}$, respectively. Each pair of roots represents a type of number with which we are already acquainted, namely (i) natural numbers, (ii) integers, (iii) rationals, (iv) reals. However, if $ax^2 + bx + c = 0$ represents each equation in turn, they all share the property that $b^2 > 4ac$.

In (v), $b^2 < 4ac$, and the result of solving this equation by using the formula of p. 116 is

$$\{x | x = 2 \pm \sqrt{(-1)}\}.$$

There is no rational or irrational number whose square is -1, and so we must either accept the position that some quadratic equations have no roots, or extend our idea of number.

Logically it would be necessary to define the field of these new numbers, laying down rules for their combination and so developing their properties. However, it is thought best at this stage merely to accept $\sqrt{(-1)}$ as a number which will obey the rules of algebra as we know them, learn how it behaves and return to strict definitions later on.

We shall therefore indicate $\sqrt{(-1)}$ by a single letter j, applying the ordinary rules of algebra, but replacing j^2 by -1 whenever it occurs.

The roots of equation (v), $x^2 - 4x + 5 = 0$, will then be written $2 - j$ and $2 + j$. Note that, if

$$x = 2 - j, \quad x^2 = 4 - 4j + j^2 = 3 - 4j,$$

and so

$$x^2 - 4x + 5 = (3 - 4j) - 4(2 - j) + 5 = 0;$$

similarly we can verify that $x = 2 + j$ is also a root of the equation. The reader should now work Exercise 8.4, p. 124, nos. 1–10.

Definitions. While not attempting to define the new numbers logically, we shall find it convenient to use certain words in a manner which will now be explained.

The numbers already known (integers, rationals, irrationals) will be called *real* numbers. Numbers such as $x + jy$, where x, y are real and $j^2 = -1$, will be called *complex* numbers.

A single symbol is often used to denote a complex number, such as $z = x + jy$.

When $y = 0$, z is said to be *real*; when $x = 0$, z is said to be a *pure imaginary* number. The two numbers $x + jy$ and $x - jy$ are called *conjugate* complex numbers.

The conjugate of z is often written $\bar{z}$, so that

$$z = x + jy, \quad \bar{z} = x - jy.$$

The number $+\sqrt{(x^2 + y^2)}$ is called the *modulus* of z, and is usually written $|z|$ or $|x + jy|$. Note that

$$\begin{aligned} z\bar{z} &= (x + jy)(x - jy) \\ &= x^2 - jxy + jxy + y^2 \\ &= (|z|)^2. \end{aligned}$$

Also

$$x = \frac{1}{2}(z + \bar{z}) \text{ and } y = \frac{1}{2j}(z - \bar{z}).$$

EXERCISE 8.4

In this exercise, use the rules of ordinary algebra but put $j^2 = -1$ *when it occurs.*

1. Show that the roots of $x^2 - 4x + 8 = 0$ are $2 - 2j$, $2 + 2j$. Verify that these complex numbers satisfy the equation.

In nos. 2–10, *write the roots of the given equations in the form* $a + jb$; *do not evaluate irrationals.*

2. $x^2 - 2x + 2 = 0$. **3.** $x^2 + 2x + 5 = 0$. **4.** $x^2 + 4x + 8 = 0$.

5. $x^2 + 2x + 4 = 0$. **6.** $x^2 - 6x + 13 = 0$. **7.** $x^2 + 4x + 9 = 0$.

8. $2x^2 - 2x + 1 = 0$. **9.** $x^2 - 3x + 3 = 0$. **10.** $9x^2 - 12x + 7 = 0$.

In nos. 11–25, *express the complex numbers in the form* $x + jy$, *where* x *and* y *are real.*

11. $(3 + 5j) + (2 - j)$. **12.** $(5 - 3j) + (-3 + 5j)$.

13. $(7 + 6j) - (3 + 2j)$. **14.** $(3 - 2j) - (4 - j)$.

15. $(2 + j)(1 - 2j)$. **16.** $(1 + j)(1 + 2j)$.

17. $(2 + j)^2$. **18.** $(1 - 2j)^2$.

19. $\dfrac{1 + j}{2 - j}$. Note that $\dfrac{1 + j}{2 - j} = \dfrac{(1 + j)(2 + j)}{(2 - j)(2 + j)}$.

20. $\dfrac{2-\mathrm{j}}{1+\mathrm{j}}$. **21.** $\dfrac{3+2\mathrm{j}}{1-2\mathrm{j}}$, **22.** $\dfrac{1}{2-3\mathrm{j}}$. **23.** $\dfrac{1}{p+\mathrm{j}q}$. **24.** $\dfrac{1}{p-\mathrm{j}q}$.

25. $\dfrac{p+\mathrm{j}q}{a+\mathrm{j}b}$.

26. Write down the real and pure imaginary parts of (i) $(3 + 4\mathrm{j})(2 + \mathrm{j})$, (ii) $\dfrac{3+4\mathrm{j}}{2+\mathrm{j}}$.

27. Write down the conjugate and the modulus of:

(i) $3 + 2\mathrm{j}$; (ii) $(2 + \mathrm{j})(-3 + 2\mathrm{j})$; (iii) $\dfrac{-1+\mathrm{j}}{2-\mathrm{j}}$.

The algebra of complex numbers

Let $z = x + \mathrm{j}y$, where x and y are real. Then

$$\begin{aligned} z = 0 &\Rightarrow z\bar{z} = 0 \\ &\Rightarrow x^2 + y^2 = 0 \\ &\Rightarrow x = 0 \text{ and } y = 0. \end{aligned}$$

Let $r = p + \mathrm{j}q$, where p and q are real. Then

$$\begin{aligned} z = r &\Rightarrow z - r = 0 \\ &\Rightarrow (x-p) + \mathrm{j}(y-q) = 0 \\ &\Rightarrow x = p \text{ and } y = q. \end{aligned}$$

If we examine the relations

$$\begin{aligned} z + r &= (x+p) + \mathrm{j}(y+q), \\ zr &= (xp - yq) + \mathrm{j}(xq + yp), \end{aligned}$$

we see that we can dispense with $\mathrm{j}^2 = -1$ once the rules for addition and multiplication have been formulated.

Moreover j itself can be dropped as it serves only to separate the real and pure imaginary parts of a complex number. z can be regarded as the *ordered number-pair* $[x, y]$, ordered because [3, 4] is different from [4, 3].

In developing the theory of complex numbers we shall now dispense with $\mathrm{j}(\mathrm{j}^2 = -1)$ although it can still be used in a practical way, if desired, to facilitate algebraic working.

A *complex number* will now be defined as an ordered pair of real numbers and represented by the symbol $[a, b]$. The rules of operation will be

(i) $[x, y] = [p, q]$ only when $x = p$ *and* $y = q$;
(ii) $[x, y] + [p, q] = [x + p, y + q]$;
(iii) $[x, y] \times [p, q] = [xp - yq, xq + yp]$.

$[x, 0]$ is called a *real complex number* and, in practice, is often written x.

$[0, y]$ is called a *pure imaginary complex number* and, in practice is often written $\mathrm{i}y$ (with $\mathrm{j}^2 = -1$).

The language of numbers with which we are familiar will be used, but it must not blind us to the fact that our numbers are complex; thus $2 + 3 = 5$ because, using rule (ii) above,

$$[2, 0] + [3, 0] = [5, 0]$$

and $2 \times 3 = 6$ because, using rule (iii) above,

$$[2, 0] \times [3, 0] = [6 - 0, 0 + 0] = [6, 0].$$

Note that $[0, 1] \times [0, 1] = [-1, 0]$. In the original symbols, $\mathrm{j}^2 = \mathrm{j} \times \mathrm{j} = -1$, a result which is meaningless in real algebra but easily comprehensible in complex algebra.

Suppose $[a, b] + [p, q] = [x, y]$; then

$$a + p = x, \quad b + q = y$$

and so

$$a = x - p, \quad b = y - q.$$

Hence

(iv) $[x, y] - [p, q] = [x - p, y - q]$, the rule for finding the difference of two complex numbers.

Again, if $[a, b] \times [p, q] = [x, y]$, then

$$ap - bq = x, \quad aq + bp = y.$$

These equations give:

$$a = (px + qy)/(p^2 + q^2), \quad b = (py - qx)/(p^2 + q^2),$$

and so

(v) $[x, y]/[p, q] = [(px + qy)/(p^2 + q^2), (py - qx)/(p^2 + q^2)]$.

This is the rule for finding the quotient of two complex numbers, but it is seldom used in practice.

EXERCISE 8.5

In nos. 1–6 *write the roots of the equations in the form* $[a, b]$, z *being a complex number.*

1. $z^2 - 4z + 13 = 0$. **2.** $z^2 - 4z - 21 = 0$. **3.** $z^2 + 4 = 0$.

4. $9z^2 + 30z + 34 = 0$. **5.** $3z^2 + 16 = 0$. **6.** $3z^2 + 4z - 14 = 0$.

In nos. 7–12 *find the sum, product, difference and quotient of the two complex numbers.*

7. $[3, -2]$, $[-1, 1]$. **8.** $[-3, 2]$, $[2, 3]$. **9.** $[3, -2]$, $[3, 2]$.

10. $[0, 4]$, $[-1, 1]$. **11.** $[2, 1]$, $[-1, 0]$. **12.** $[-2, 5]$, $[2, -5]$.

In nos. 13–18 *a complex number* z *is given; write down* z_1 *and* z_2 *so that* $z + z_1 = 0$, $zz_2 = [1, 0]$.

13. $[1, 1]$. **14.** $[2, -1]$. **15.** $[-2, 1]$. **16.** $[0, 1]$. **17.** $[-1, 0]$.
18. $[3, -4]$.

19. If z_1, z_2, z_3 are three complex numbers, verify that:
(i) $z_1z_2 = z_2z_1$; (ii) $z_1 + z_2 = z_2 + z_1$; (iii) $(z_1 + z_2) + z_3 = z_1 + (z_2 + z_3)$; (iv) $(z_1z_2)z_3 = z_1(z_2z_3)$; (v) $z_1(z_2 + z_3) = z_1z_2 + z_1z_3$.

CHAPTER 9

SEQUENCES, SERIES, LIMITS

Sequences

A *sequence* is an ordered set of numbers in which the individual numbers, or *terms*, are decided by their positions in the order. For example, the sequence

$$4, 8, 12, 16, \ldots$$

can be formed from the set of natural numbers 1, 2, 3, 4, ... by multiplying each member by 4.

The terms of a sequence are commonly denoted by the letter a with an appropriate suffix: thus a sequence can be written $a_1, a_2, a_3, a_4, \ldots$. The *general term* of the sequence is written a_r.

Hence we can define a sequence $\{a_r\}$ as a function having the ordered set of natural numbers $\{r\}$ as its domain.

Examples of sequences will now be given.

(i) $1, 3, 5, 7, \ldots,$ in which $a_r = 2r - 1$;

(ii) $1, \frac{1}{2}, \frac{1}{4}, \frac{1}{8}, \ldots,$ in which $a_r = 1/2^{r-1}$;

(iii) $-1, 4, -27, 256, \ldots,$ in which $a_r = (-r)^r$;

(iv) $0, \frac{1}{2}, \frac{3}{4}, \frac{7}{8}, \ldots,$ in which $a_r = 1 - 1/2^{r-1}$.

EXERCISE 9.1

In nos. 1–12, *r is a positive integer; write down the first five terms of the sequences whose general term* a_r *is given:*

1. $2r$. **2.** $3r + 5$. **3.** 2^r. **4.** $12/3^{r-1}$. **5.** $(-3)^r$.
6. $4 - (-1)^r$. **7.** $(-r)^3$. **8.** $r - 1/r$. **9.** $r(r + 1)$. **10.** $2r + (-1)^r$.
11. $r^2 - (r - 1)^2$. **12.** $r!$ (See p. 142.)

13. Write down the first five terms of the sequence $\{a_r\}$ in which $a_1 = 1$, $a_2 = 1$, $a_r = a_{r-1} + a_{r-2}$.

In nos. 14–22 *give the next two terms of the sequence and the general term* a_r:

14. $1, 4, 7, 10, \ldots$ **15.** $1, 4, 16, 64, \ldots$ **16.** $1, -\frac{1}{2}, \frac{1}{3}, -\frac{1}{4}, \ldots$

17. $5, 3, 1, -1, \ldots$ **18.** $2, -4, 8, -16, \ldots$ **19.** $2^2, 4^2, 6^2, 8^2, \ldots$

20. $1, 3, 7, 15, \ldots$ **21.** $0, -\frac{1}{2}, -\frac{3}{4}, -\frac{7}{8}, \ldots$ **22.** $0, 1\cdot5, 2\cdot\dot{6}, 3\cdot75, \ldots$

Note. It is dangerous to assume that the function which generates a sequence can be uniquely determined from the first few terms. For example,

$$1, 3, 5, 7, \ldots.$$

would be the first four terms generated both by $a_r = 2r - 1$ and by $a_r = (2r - 1) + \frac{1}{5}(r - 1)(r - 2)(r - 3)(r - 4)$. Clearly many other forms of a_r would give the same first four terms of the sequence $\{a_r\}$.

The arithmetic sequence (A.S.)

A sequence of the form

$$a, \quad a + d, \quad a + 2d, \quad a + 3d, \ldots$$

where a and d are constants which may be positive or negative, is called an *arithmetic sequence*, often abbreviated to A.S. It is also known as an arithmetical progression (A.P.).

The constant a is the first term, and the constant d, since it is always added to a term to produce the next term, is called the *constant difference*.

Writing the terms of the A.S. in the form

$$a + 0 \,.\, d, \quad a + 1 \,.\, d, \quad a + 2 \,.\, d, \ldots$$

we have $a_r = a + (r - 1)\,d$. If there are n terms in the sequence, the last term, a_n, is $a + (n - 1)\,d$.

The set of the first n positive integers is a most important special case of an A.S. Suppose the sum of this set is S_n, that is,

$$S_n = 1 + 2 + 3 + \ldots + n.$$

Reversing the order of the terms,

$$S_n = n + (n - 1) + (n - 2) + \ldots + 1.$$

Adding the two forms of S_n,

$$2S_n = (n + 1) + (n + 1) + (n + 1) + \ldots + (n + 1)$$
$$= n(n + 1), \quad \text{since there are } n \text{ terms.}$$

Hence

$$S_n = \tfrac{1}{2}n(n + 1).$$

E. T. Bell relates the story, in his *Men of Mathematics*, of the discovery of this method of evaluating S_n by Gauss, while still a young boy at school.

The sum of the general arithmetic sequence can now be deduced; thus,

$$S_n = a + (a + d) + (a + 2d) + \ldots + \{a + (n - 1)\,d\}$$
$$= na + d\{1 + 2 + 3 + \ldots + (n - 1)\}$$
$$= na + \tfrac{1}{2}dn(n - 1),$$

or

$$\mathbf{S}_n = \tfrac{1}{2}\boldsymbol{n}\{\mathbf{2}\boldsymbol{a} + (\boldsymbol{n} - \mathbf{1})\boldsymbol{d}\}.$$

Ex. 1. Find the nth term of the A.S.

$$1, 5, 9, 13, \ldots.$$

Find also the sum of n terms of this A.S.

Ex. 2. If a_1, a_n are the first and last terms of an A.S. of n terms, and S_n is the sum of the terms, show that $\mathbf{S}_n = \tfrac{1}{2}\boldsymbol{n}(\boldsymbol{a}_1 + \boldsymbol{a}_n)$.

The arithmetic mean

The *arithmetic mean* of the n numbers $a_1, a_2, a_3, \ldots a_n$ is defined as

$$\frac{a_1 + a_2 + a_3 + \ldots + a_n}{n}.$$

The arithmetic mean is also commonly called the *average*.

The arithmetic mean of the terms of the A.S. $\{a_1, a_2, a_3, \ldots a_n\}$ is S_n/n, or $\tfrac{1}{2}\{2a + (n - 1)\,d\}$.

A convenient way of writing three terms which form an A.S. is $(a - d)$, a, $(a + d)$; if there are four terms we write $(a - 3d)$, $(a - d)$, $(a + d)$, $(a + 3d)$.

Ex. 3. Show that the arithmetic mean of all the terms of an A.S. is the same as the arithmetic mean of the first and last terms.

Ex. 4. The first and fourth terms of an A.S. are 11 and 3; write down the two intervening terms.

EXAMPLE. *A man's salary starts at* \$180 *a month, and increases after each period of six months by* \$12 *a month. Find his salary in the eighth month of his third year, and the total amount he receives in six years.*

His salary in dollars at six-monthly periods forms the sequence 180, 192, 204, ... His salary in the eighth month of the third year is the sixth term of this sequence, that is,

$$180 + 5 \,.\, 12,$$

and so is \$240.

The sums received in dollars over successive 6-monthly periods form the A.S.

$$1080, \quad 1152, \quad 1224, \ldots .$$

Hence the total amount received in six years is

$$\$\tfrac{1}{2} \times 12 \times \{2160 + 11 \times 72\} = \$17{,}712.$$

EXAMPLE. *Find how many terms of the* A.S. 3, 7, 11, . . . *are needed for their sum to exceed* 260.

Suppose the sum of n terms of the A.S. 3, 7, 11, . . . is exactly 260; then

$$\tfrac{1}{2}n\{6 + 4(n - 1)\} = 260$$

so

$$2n^2 + n - 260 = 0.$$

The positive root of this quadratic equation is approximately 11·16.

Hence 12 terms of the A.S. are needed for their sum to exceed 260.

Ex. 5. For the A.S. 83, 77, 71, . . ., find (i) the first negative term, and (ii) after how many terms the sum is negative.

The geometric sequence (G.S.)

A sequence of the form $a, ac, ac^2, \ldots$ in which each term is formed by multiplying the preceding term by a constant, c, is called a *geometric sequence*, written G.S. It is also called a geometric progression (G.P.).

Suppose a_n is the nth term of the G.S. $a, ac, ac^2, \ldots$. If we write the terms in the form $ac^0, ac^1, ac^2, \ldots$ we see that

$$a_n = ac^{n-1}.$$

The multiplying factor, c, is called the *constant ratio.*

Suppose S_n is the sum of the first n terms of the G.S. $a, ac, ac^2, \ldots$. Then

$$S_n = a + ac + ac^2 + \ldots + ac^{n-2} + ac^{n-1},$$

so

$$cS_n = \quad ac + ac^2 + \ldots \quad + ac^{n-1} + ac^n;$$

subtracting,

$$S_n(1-c) = a - ac^n \quad \text{or} \quad S_n(c-1) = ac^n - a.$$

Hence

$$S_n = \frac{a(1-c^n)}{1-c}, \quad \text{or} \quad S_n = \frac{a(c^n-1)}{c-1}.$$

When $0 < c < 1$,

$$S_n = \frac{a}{1-c} - \frac{ac^n}{1-c},$$

the first term being independent of n. In the second term there is the factor c^n, which gets steadily smaller as n gets larger. The value of S_n therefore differs from $a/(1-c)$ by a number which approaches zero as n increases.

The geometric mean

The *geometric mean* of the n numbers $a_1, a_2, a_3, \ldots a_n$ is the nth root of the product of the numbers, that is, $(a_1a_2a_3 \ldots a_n)^{1/n}$.

The geometric mean of a and b is $\sqrt{(ab)}$, and of a, ac, ac^2 is $\sqrt[3]{(a^3c^3)}$, or ac.

Ex. 6. Show that $-1, 2, -4, 8$ form a G.S. What is the constant ratio? What is their geometric mean?

Ex. 7. Insert four terms between 8 and $\frac{1}{4}$ so that the six terms form a G.S.

EXAMPLE. *A crab advances* 1 *metre, then* $\frac{1}{2}$ *m, then* $\frac{1}{4}$ *m, and so on, each move being one-half of the preceding move. Find the total distance he has covered after* (*i*) 5 *moves,* (*ii*) *n moves.*

If he were able to continue indefinitely in this way, is there any limit to the total distance covered?

The total distance covered after n moves is the sum of the terms of a G.S.:

$$1 + \frac{1}{2} + \frac{1}{4} + \ldots + \frac{1}{2^{n-1}} \text{ metres.}$$

(i) After 5 moves the distance is

$$\frac{1 - (\frac{1}{2})^5}{1 - \frac{1}{2}} = 2 - \frac{1}{2^4} = 1{\cdot}9375 \text{ metres.}$$

(ii) After n moves the distance is

$$\frac{1 - (\frac{1}{2})^n}{1 - \frac{1}{2}} = 2 - \frac{1}{2^{n-1}} \text{ metres.}$$

However large n may be, the distance covered cannot reach a total of 2 metres.

Note. Suppose that, in the nth move, the crab covers $1/n$ metres. The total distance in metres covered in n moves is then

$$\begin{aligned} & 1 + \tfrac{1}{2} + \tfrac{1}{3} + \tfrac{1}{4} + \tfrac{1}{5} + \ldots, \\ = \; & 1 + \tfrac{1}{2} + (\tfrac{1}{3} + \tfrac{1}{4}) + (\tfrac{1}{5} + \tfrac{1}{6} + \tfrac{1}{7} + \tfrac{1}{8}) + \ldots, \\ > \; & 1 + \tfrac{1}{2} + 2(\tfrac{1}{4}) + 4(\tfrac{1}{8}) + 8(\tfrac{1}{16}) + \ldots, \\ = \; & 1 + \tfrac{1}{2} + \tfrac{1}{2} + \tfrac{1}{2} + \tfrac{1}{2} + \ldots. \end{aligned}$$

Hence, in 2^n moves, the distance covered is greater than $(1 + \frac{1}{2}n)$ metres, if $n > 2$. In this case there would be no limit to the distance the crab could cover if he were able to continue indefinitely.

EXAMPLE. *Find the sum of* (*i*) 19, (*ii*) 20, (*iii*) *n terms of the sequence* 48, −24, +12, −6, ...

What happens to the sum of n terms as n increases?

The given sequence is a G.S.:

$$48, \quad 48(-\tfrac{1}{2}), \quad 48(-\tfrac{1}{2})^2, \ldots .$$

(i) The sum of 19 terms is

$$S_{19} = \frac{48\{1-(-\frac{1}{2})^{19}\}}{1-(-\frac{1}{2})} = 32 + \frac{1}{2^{14}}.$$

(ii) The sum of 20 terms is

$$S_{20} = \frac{48\{1-(-\frac{1}{2})^{20}\}}{1-(-\frac{1}{2})} = 32 - \frac{1}{2^{15}}.$$

(iii) The sum of n terms is

$$S_n = \frac{48\{1-(-\frac{1}{2})^{n}\}}{1-(-\frac{1}{2})} = 32 - (-1)^n \frac{1}{2^{n-5}}.$$

As n increases, S_n approaches more and more closely to 32, but it *oscillates* on either side of 32 since the term $(-1)^n 1/2^{n-5}$ is alternately negative and positive.

Ex. 8. Find the sum of (i) 9, (ii) 10, (iii) $2n$, (iv) $2n+1$ terms of the G.S. $\frac{1}{9}, -\frac{1}{6}, \frac{1}{4}, - \ldots$

EXERCISE 9.2

In nos. 1–9 *say whether the sequence could be an* A.S., *a* G.S., *or neither. Write down the nth term of the sequence and, if it is either an* A.S. *or a* G.S., *give the sum of the first n terms of the sequence.*

1. 4, 7, 10, . . . **2.** 12, 6, 3, . . . **3.** $1^2, 2^2, 3^2, \ldots$

4. 1, 3, 9, . . . **5.** 7, 3, −1, . . . **6.** −2, 4, −8, . . .

7. $\frac{1}{48}, \frac{1}{24}, \frac{1}{12}, \ldots$ **8.** $\frac{1}{48}, \frac{1}{24}, \frac{1}{16}, \ldots$ **9.** $\frac{1}{2}, \frac{1}{5}, \frac{1}{8}, \ldots$

10. Find the sum of the positive integers from 1 to 100 excluding all multiples of 3.

11. By how much does the sum of the first 59 even positive integers exceed the sum of the first 59 odd integers?

12. The positive integers are grouped as follows:

$$(1) + (2 + 3) + (4 + 5 + 6) + \ldots$$

(i) How many integers are included in the first 7 groups? (ii) What is the first integer and the sum of the integers in the eighth group?

13. Find the sum of:
(i) $1 - 2 + 3 - 4 + \ldots - 2r + (2r + 1)$;
(ii) $1^2 - 2^2 + 3^2 - 4^2 + \ldots - (2r)^2 + (2r + 1)^2$.

14. Find the first three terms of the A.S. whose:
(i) 4th term is 25 and 9th term is -10; (ii) 7th term is $3a + 5b$ and 18th term is $19(2b - a)$.

15. Find the fifth term of an A.S. when:
(i) the 4th term is 5 and the sum of the first 10 terms is 70; (ii) the 3rd term is $(3x + 6)$ and the sum of the first 8 terms is $4(8x + 21)$.

16. Find the first three terms of a G.S. whose:
(i) 4th term is 40 and 8th term is 2·5; (ii) 2nd term is -5 and 7th term is 1215.

17. If a and b are two unequal positive numbers, prove that the arithmetic mean of a and b is greater than their geometric mean.

18. Find how many terms of the following sequences must be taken for the sum of the terms to exceed 100:
(i) 1, 5, 9, . . .; (ii) 1, 2, 4, 8, . . .; (iii) 7, 8·5, 10, . . .;
(iv) -3, 9, -27, . . .; (v) 52, 26, 13, . . .; (vi) -3, 9, -15, 21, . . .

19. If \$850 is invested at $2\frac{1}{2}$ per cent compound interest, find, to the nearest dollar, its value at the end of (i) one year, (ii) ten years, to the nearest dollar.

20. If the value of a machine, originally \$4,000, depreciates each year by 15 per cent of its value at the beginning of the year, find its value after 3 years, and after 7 years, to the nearest dollar.

21. When his first child was born a man invested £50 in a deposit account in which compound interest at 3 per cent was added annually, and he continued to invest £50 in the same account on each of the child's subsequent birthdays. Find the amount in the account just before the 21st birthday, to the nearest £1.

22. A loan of \$15,000 is made with interest at 5 per cent per annum. It is repaid by annual payments of \$2,000, starting at the end of the first year. Find the amount outstanding, to the nearest \$100, at the end of the 8th year.

The geometric series

We have seen that, when $c < 1$,

$$\frac{1 - c^n}{1 - c} = 1 + c + c^2 + \ldots + c^{n-1}.$$

Although this is strictly an algebraic identity, since we can show that $(1 - c)(1 + c + c^2 + \ldots + c^{n-1}) \equiv 1 - c^n$, it does suggest the idea that a function can be represented by a *series*, that is, by adding together the terms of a sequence.

Suppose

$$S_n = 1 - \frac{1}{2} + \frac{1}{4} - \frac{1}{8} + \ldots + \frac{(-1)^{n-1}}{2^{n-1}}.$$

In Fig. 9.1 the values of S_1, S_2, S_3, S_4 are represented by the lengths OA_1, OA_2, OA_3, OA_4.

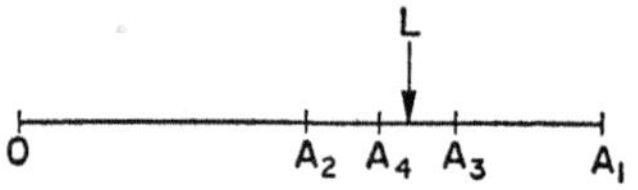

Fig. 9.1

Now

$$S_n = \{1 - (-\tfrac{1}{2})^n\}/\{1 - (-\tfrac{1}{2})\} = \tfrac{2}{3} - \tfrac{2}{3}(-\tfrac{1}{2})^n.$$

Thus

$$OA_1 = \tfrac{2}{3} + \tfrac{1}{3},\ OA_2 = \tfrac{2}{3} - \tfrac{1}{6},\ OA_3 = \tfrac{2}{3} + \tfrac{1}{12},\ OA_4 = \tfrac{2}{3} - \tfrac{1}{24},$$

and it is not difficult to visualize the continuation of this process. If $OL = \tfrac{2}{3}$, successive points A_r and A_{r+1} will lie on opposite sides of L, and the distance $LA_r = \tfrac{2}{3}(\tfrac{1}{2})^r$ will get steadily smaller as r increases.

Furthermore, we can make the distance LA_r as small as we please by making r sufficiently large.

For example, let

$$LA_r < 0{\cdot}001,$$

so

$$\tfrac{2}{3}(\tfrac{1}{2})^r < 0{\cdot}001,$$

i.e.

$$2^{r-1} > 1/0{\cdot}003,$$

thus

$$r > 1 + \frac{3 - \log 3}{\log 2}.$$

This gives $r > 9{\cdot}38$ (approximately), and so $LA_r < 0{\cdot}001$ when $r \geqslant 10$.

To sum up, we can say that $|S_n - \frac{2}{3}| < 0{\cdot}001$ when $n > 9$.

Ex. 9. Find the value of N such that $|S_n - \frac{2}{3}| < 0{\cdot}0001$ when $n > N$. Repeat for $|S_n - \frac{2}{3}| < 0{\cdot}00001$.

Thus, for this series, S_n can be made as near to $\frac{2}{3}$ as we like, by taking n sufficiently large.

The value of $\frac{2}{3}$ can be expressed as a recurring decimal, as we saw in the consideration of nested intervals in Chapter 1. We now see that $\frac{2}{3}$ could also be evaluated as a decimal, to as high a degree of accuracy as we desire, by evaluating S_n for a sufficiently large value of n, where

$$S_n = 1 - \frac{1}{2} + \frac{1}{4} - \frac{1}{8} + \ldots + \frac{(-1)^{n-1}}{2^{n-1}}.$$

Limit of a sequence

In the last section we considered the values of the sequence $S_1, S_2, S_3, \ldots S_n$, particularly for increasingly large values of n. We saw that, when we had chosen a number ϵ, however small, we could make $|S_n - \frac{2}{3}| < \epsilon$ by taking a sufficiently large value of n. In several cases we found a number N, depending on ϵ, such that $|S_n - \frac{2}{3}| < \epsilon$ when $n > N$; in particular, when $\epsilon = 0{\cdot}001$, we found that $N = 9$.

Figure 9.1 gave a picture of how the particular sequence S_r which we were considering, settled more and more nearly to the value $\frac{2}{3}$. We express the result briefly by saying that the limit of S_n as n tends to infinity is $\frac{2}{3}$; even more briefly we write

$$\lim_{n \to \infty} S_n = \frac{2}{3}.$$

In this work n is a positive integer, and the term S_n of the sequence $\{S_r\}$ is a known function of n. A general definition of the limit of a sequence will now be given.

Definition. $f(1), f(2), f(3), \ldots f(n)$ are the terms of a sequence $\{f(r)\}$. Suppose there is a constant number l which satisfies the following conditions: when we choose a number ϵ, however small, we can find a number N, depending on ϵ, such that $|f(n) - l| < \epsilon$ when $n > N$; then $f(n)$ approaches the limit l as n tends to infinity.

We write $f(n) \to l$ as $n \to \infty$, or

$$\lim_{n\to\infty} f(n) = l.$$

Notes.

(i) The limiting value of a sequence is not, in general, a term of the sequence itself. For example, if $f(n) = 1 - 2^{-n}$, then $\lim_{n\to\infty} f(n) = 1$ but there is no value of n for which $f(n) = 1$.

(ii) The use of the symbol $\exists$ enables us to abbreviate the definition of a limit. We write:

$$\epsilon > 0;\ \exists\ N : n > N \Rightarrow |f(n) - l| < \epsilon,$$

then

$$\lim_{n\to\infty} f(n) = l.$$

Ex. 10. Find the limit of the sequence $\{f(n)\}$ as $n \to \infty$ when $f(n)$ is:

(i) $\dfrac{n}{n+1}$; (ii) $\dfrac{n(n+2)}{2n^2+1}$; (iii) $\dfrac{2^n+3}{2^{n-1}}$.

Limit of a function

Figure 9.2 represents a portion of the graph of $y = f(x)$. A is a point on OX such that $OA = a$. The *neighbourhood* of A on OX is the set of all points P, where $OP = x$, such that $|x - a| < \delta$, where δ is a positive number. More strictly this set is a δ-neighbourhood, and in what follows we shall be thinking of small values of δ.

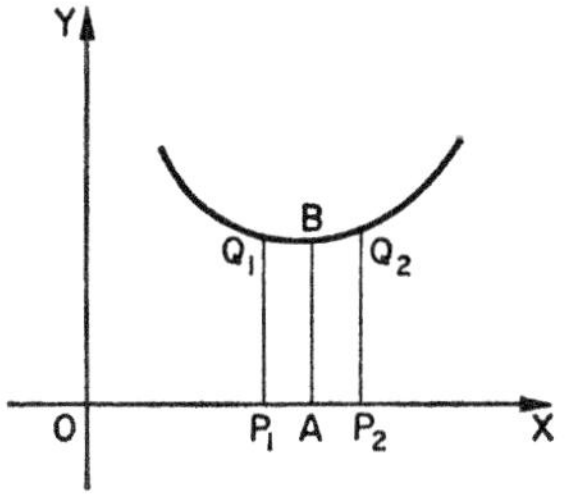

Fig. 9.2

The idea of a limiting value of the function $f(x)$ as x approaches the value a arises naturally from the idea of the limiting value of a sequence. The limit exists if there is a fixed number l such that we can make $f(x)$ as near as we wish to l by taking x sufficiently near to a, that is, by placing x in a suitable neighbourhood of a.

Ex. 11. Prove that $(x + 1/x)$ differs from 2 by less than 0·1 provided that x differs from 1 by less than 0·26.

The function $(x + 1/x)$ approaches the limit 2 as x approaches the value of 1 and, in this case, it is also true that the function has the value 2 when $x = 1$. However, in general, the consideration of $\lim_{x \to a} f(x)$ says nothing whatever about $f(a)$ itself; it is not even implied that $f(a)$ exists. For example, it is shown in Chapter 22 that $\lim_{x \to 0} (\sin x/x) = 1$, but the function $\sin x/x$ does not exist when $x = 0$.

Formal definition of $\lim_{x \to a} f(x)$

The limit of $f(x)$ as x tends to a exists, and is a fixed number l, provided that, when we have chosen a positive number ϵ, however small, we can always find a suitable δ-neighbourhood of a within which $|f(x) - l| < \epsilon$.

More briefly, $\lim_{x \to a} f(x) = l$ provided that

$$\epsilon > 0;\ \exists\delta(\epsilon): |x - a| < \delta \Rightarrow |f(x) - l| < \epsilon.$$

Ex. 12. Show that, when $\epsilon > 0$,

$$\left|\frac{x}{x^2+1}\right| < \epsilon$$

when $|x| < \epsilon$. What is $\lim_{x\to 0} \{x/(x^2+1)\}$?

EXERCISE 9.3

1. S_n represents the sum of n terms of the following geometric series. Express S_n in terms of n and find $\lim_{n\to\infty} S_n$.

(i) $1 + \frac{1}{2} + \frac{1}{4} + \ldots + \frac{1}{2^{n-1}}$;

(ii) $3 - 1 + \frac{1}{3} - \ldots + \frac{(-1)^{n-1}}{3^{n-2}}$;

(iii) $1 + \frac{2}{3} + \frac{4}{9} + \ldots + \left(\frac{2}{3}\right)^{n-1}$.

2. Write down the limits, as $n \to \infty$, of the sequences whose nth terms are:

(i) $\frac{2n}{3n+4}$; (ii) $\frac{3n-2}{4n+3}$; (iii) $\frac{n+1}{2n^2+1}$;

(iv) $\frac{2^n+1}{3^n+2}$; (v) $\frac{3^n-2^n}{3^n+2^n}$; (vi) $\frac{2^{n+1}-3^n}{2^n-3^{n+1}}$.

3. Find a suitable neighbourhood of 2 within which $\left|\left(x+\frac{2}{x}\right)-3\right| < 0{\cdot}1$.

What is $\lim_{x\to 2}\left(x+\frac{2}{x}\right)$?

4. Show that $\left|\frac{x-1}{x^2-1}-\frac{1}{2}\right| < 0{\cdot}3$ when $0{\cdot}25 < x < 4$. What is $\lim_{x\to 1}\left(\frac{x-1}{x^2-1}\right)$?

5. Find the neighbourhood of $x = 1$ for which

$$\left|\frac{x-1}{x^2-1}-\frac{1}{2}\right| < \epsilon.$$

6. Find the limit as $n \to \infty$ of $\sqrt{(n+1)} - \sqrt{n}$.

CHAPTER 10

MATHEMATICAL INDUCTION AND APPLICATIONS

The principle of mathematical induction

Consider the sequence $\{s_r\}$, where

$$s_1 = 1, \quad s_2 = 1 + 3, \quad s_3 = 1 + 3 + 5, \quad s_4 = 1 + 3 + 5 + 7,$$

and so on. The exact meaning of the phrase "and so on" is not obvious unless we can write down a formula for s_n; in this case

$$s_n = 1 + 3 + 5 + \ldots + (2n - 1).$$

Closer examination of the expression for s_n shows there is still a gap which is not covered explicitly; the three dots ... indicate the terms between the third and the nth.

We can get another view of the sequence $\{s_r\}$ by evaluating successive terms:

$$s_1 = 1, \quad s_2 = 4 = 2^2, \quad s_3 = 9 = 3^2, \quad s_4 = 16 = 4^2.$$

It would appear that if "and so on" is correctly applied, we should have $s_5 = 5^2, s_6 = 6^2, \ldots, s_n = n^2$.

The gap indicated by three dots ... still shows the lack of a definite formula, and we cannot do more than say that $s_n = n^2$ is very probably true. It is the purpose of the method of *mathematical induction* to bridge this gap. Briefly this method would show that, if $s_n = n^2$ is true, then $s_{n+1} = (n + 1)^2$ is also true; but $s_1 = 1^2$ is true, and so $s_n = n^2$ is true for all values of n.

s_n is changed into s_{n+1} by adding an extra term $2n + 1$. Then $s_n = n^2$ implies that $s_{n+1} = n^2 + (2n + 1) = (n + 1)^2$. Assuming the principle of mathematical induction, the proof would now be complete, and $s_n = n^2$ for all n.

It is important to avoid confusion with the meaning of the word "induction" as applied in science and logic, where it indicates the formulation of a postulate after the examination of particular cases, and the truth of the postulate is tested for failure by deduction and further experiment.

Before establishing the principle of mathematical induction we shall consider the purely inductive part, where a property of the general term of a sequence is first postulated.

Ex. 1. Suggest a formula for s_n when:

(i) $s_n = \dfrac{1}{1.2}+\dfrac{1}{2.3}+\dfrac{1}{3.4}+\ldots+\dfrac{1}{n(n+1)}$;

(ii) $s_n = 1+7+19+\ldots+(3n^2-3n+1)$;

(iii) $s_n = 1^3+2^3+3^3+\ldots+n^3$.

Ex. 2. Find the highest common factor of the terms of $\{s_n\}$ when:
(i) $s_n = 5^{2n}+3n-1$; (ii) $9(9^n-1)-8n$.

Ex. 3. Which is eventually the greater, 2^{2n} or $n!$?
[$n!$ denotes the product $1.2.3\ldots n$; for example, $4! = 1.2.3.4 = 24$.]

Proof of the principle

The axiom on which the proof of the principle of mathematical induction will be based is:

Any non-empty set N of positive integers must contain a smallest member.

More precisely, there must be a member l of N such that

$$n \in N \Rightarrow l \leqslant n.$$

We next prove that there cannot be an integer between 0 and 1. This seems obvious enough, but there is little point in adding to the list of axioms something that can very easily be proved. Suppose the statement is wrong, that is, suppose c is an integer and $0 < c < 1$. Then c belongs to the set C of positive integers, which is not an empty set. Let l be the least member of C, so that, if c exists, $0 < l < 1$; then $0 < l^2 < l$; this contradicts the hypothesis that l is the least member of C. A hypothesis that leads by deduction to a contradiction must be false (a form of argument known as *reductio ad absurdum*), and so there cannot be an integer between 0 and 1.

The next theorem contains the essence of the principle of mathematical induction:

A set N of positive integers which includes 1, *and which includes* $(n + 1)$ *whenever it includes n, includes every positive integer.*

If N does not contain all the positive integers, let N' contain all positive integers not contained in N; we suppose that N' is not empty. Let the least element of N' be l. Since 1 belongs to N and there is no integer between 0 and 1, it follows that $l > 1$. Hence $l - 1 > 0$; but $l - 1 < l$, so that $l - 1 \in N$.

The assumed property of the set N ensures that $(l - 1) + 1 \in N$, or $l \in N$. We now have $l \in N$, $l \in N'$, $N \cap N' = \Phi$. The supposition that $N' \neq \Phi$ thus leads to a contradiction, and so must be false. Hence all the positive integers are members of N.

Finally, let us consider a statement $P(n)$ associated with each positive integer n. $P(n)$ represents a proposition that is true; for example, $P(n)$ could stand for the true proposition

$$1 + 2 + 3 + \ldots + n = \tfrac{1}{2}n(n + 1).$$

Suppose that $P(1)$, and that, for all n, $P(n) \Rightarrow P(n + 1)$. By this we mean that the proposition is true when $n = 1$, and that, if the proposition is true for the integer n, it is also true for the integer $(n + 1)$. Then the set of positive integers for which $P(n)$ applies contains 1 and also, if it contains n, it must also contain $(n + 1)$; it therefore contains all the positive integers. Hence $P(n)$ represents a true statement for all positive integral values of n.

This concludes the formal proof of the principle of mathematical induction; it can be briefly stated as follows:

If $P(1)$, and $P(n) \Rightarrow P(n + 1)$, then $P(n)$ for all n.

In this statement n is a positive integer.

EXAMPLE. *Prove that the sum of the first n odd integers is equal to* n^2.

Let $s_1 = 1$, $s_2 = 1 + 3, \dots s_n = 1 + 3 + 5 + \dots + (2n - 1)$. If $P(n)$ denotes the statement $s_n = n^2$, then $P(1)$ is true.

Assume $P(n)$, so that $s_n = n^2$. But

$$\begin{aligned} s_{n+1} &= s_n + (2n + 1) \\ &= n^2 + 2n + 1 \\ &= (n + 1)^2. \end{aligned}$$

Hence $P(1)$ and $P(n) \Rightarrow P(n + 1)$. The result now follows from the principle of mathematical induction.

Ex. 4. Prove by induction the formulae for the sum of n terms of an A.S. and of a G.S.

EXAMPLE. *Prove that* $9(9^n - 1) - 8n$ *is divisible by* 64.

Let $f(n) = 9(9^n - 1) - 8n$, so that $f(1) = 64$.

[If a and b are positive integers, the statement a *is divisible exactly by* b is written $b|a$.]

If $P(n)$ denotes the statement $64|f(n)$, then $P(1)$ is true.

Assume $P(n)$, so that $64|f(n)$. But

$$\begin{aligned} f(n + 1) - 9f(n) &= \{9^{n+2} - 9 - 8(n + 1)\} - 9\{9^{n+1} - 9 - 8n\} \\ &= -9 - 8 + 81 \\ &= 64, \end{aligned}$$

so that $f(n + 1) = 9f(n) + 64$.

Hence $P(1)$ and $P(n) \Rightarrow P(n + 1)$, and the result follows from the principle of mathematical induction.

Ex. 5. Prove that $5^{2n} + 3n - 1$ is divisible by 9.

EXAMPLE. *Prove that* $n! > 3^n$ *when* $n \geqslant 7$.

Let $P(n)$ denote the statement $n! > 3^n$. Evaluation of $n!$ and 3^n for $n = 1, 2, 3, 4, 5, 6$ shows that $P(n)$ is false for all these values n, but $7! = 5040$, $3^7 = 2187$ and so $P(7)$ is true.

Assume $P(n)$ when $n \geqslant 7$, so that

$$n! > 3^n \text{ when } n \geqslant 7.$$

Hence

$$(n+1)\,n! > (n+1)\,.\,3^n$$
$$> 3\,.\,3^n, \quad \text{since } n+1 > 3,$$

giving

$$(n+1)! > 3^{n+1} \quad \text{when } n \geqslant 7.$$

Therefore $P(7)$ and $P(n) \Rightarrow P(n+1)$ when $n \geqslant 7$, and the result follows from the principle of mathematical induction.

Note that this Example does not fall exactly into line with the proof of the principle as given on pp. 142–3. The reader will not find it difficult to modify the proof to establish that, if $P(k)$ and $P(n) \Rightarrow P(n+1)$ when $n \geqslant k$, then $P(n)$ when $n \geqslant k$.

EXAMPLE. *$\{u_n\}$ is a sequence determined by $u_0 = 2$, $u_1 = 3$, $u_{n+1} = 3u_n - 2u_{n-1}$. Prove that $u_n = 2^n + 1$.*

Let $P(n)$ denote the statement $u_n = 2^n + 1$. Then $P(0)$ and $P(1)$. Assume $P(n-1)$ and $P(n)$, so that

$$u_{n-1} = 2^{n-1} + 1 \quad \text{and} \quad u_n = 2^n + 1.$$

It follows that

$$u_{n+1} = 3(2^n + 1) - 2(2^{n-1} + 1)$$
$$= 2\,.\,2^n + 1$$
$$= 2^{n+1} + 1.$$

Hence $P(n-1)$ *and* $P(n) \Rightarrow P(n+1)$, but $P(0)$ *and* $P(1)$, so the result follows from the principle of mathematical induction.

Again the established form of the principle has not been followed exactly, but the necessary extension of the proof is not difficult. In more difficult cases the reader may find it necessary to assume $P(k)$ for all values of k from 1 to n in order to establish $P(n+1)$; this again is a valid extension of the principle. For a proof of this statement the reader should consult *A Brief Survey of Modern Algebra* by Birkhoff and MacLane (Macmillan), p. 13.

A warning

In using the principle of mathematical induction it is important to understand that:

(i) the principle can only apply to statements involving positive integers;

(ii) the sequence of correct statements must have a definite beginning, usually with $P(1)$.

As far as (i) is concerned, a glance at the proof of the principle on pp. 142–3 will be convincing.

In the case of (ii) the well-known example which follows will provide both conviction and amusement.

Prove that $n > (n + 1)$.

Let $P(n)$ denote the statement $n > (n + 1)$.

Assume $P(n)$, so that

$$n > (n + 1),$$

and so

$$n + 1 > (n + 1) + 1,$$

or

$$n + 1 > n + 2,$$

which gives $P(n + 1)$. Hence $P(n) \Rightarrow P(n + 1)$, and so by the principle of mathematical induction . . . !

EXERCISE 10.1

In nos. 1–15 *use the principle of mathematical induction to prove the statement made:*

1. $1.2 + 2.3 + 3.4 + \ldots + n(n + 1) = \frac{1}{3}n(n + 1)(n + 2)$.

2. $\frac{1}{1.2} + \frac{1}{2.3} + \frac{1}{3.4} + \ldots + \frac{1}{n(n + 1)} + \frac{n}{n + 1}$.

3. $1^2 + 2^2 + 3^2 + \ldots + n^2 = \frac{1}{6}n(n + 1)(2n + 1)$.

4. $1^3 + 2^3 + 3^3 + \ldots + n^3 = \frac{1}{4}n^2(n + 1)^2$.

5. $2 + 2^2 + 2^3 + \ldots + 2^n = 2(2^n - 1)$.

6. $2^n > n$.

7. $n! > 4^n$ when $n \geqslant k$, the value of k to be determined.

8. $2^n > n^2$ when $n = 1$ and when $n > 4$.

9. $(1 + \alpha)^n > 1 + n\alpha$ when $1 + \alpha > 0$, $\alpha \neq 0$, $n > 1$.

10. $\frac{1}{\sqrt{1}} + \frac{1}{\sqrt{2}} + \frac{1}{\sqrt{3}} + \ldots + \frac{1}{\sqrt{n}} > \sqrt{n}$ when $n > 1$.

11. $9^n + 7$ is divisible by 8.

12. $(3n + 1) . 7^n - 1$ is divisible by 9.

13. $11^{n+2} + 12^{2n+1}$ is divisible by 133.

14. $7^{2n} - 48n - 1$ is divisible by 2304 when $n > 1$.

15. The sum of the cubes of any three consecutive natural numbers is divisible by 9.

16. What common factor occurs in all numbers of the form (i) $(2r - 1)^2 - 1$, (ii) $(2r - 1)^2 + 1$, when r is a positive integer?

17. Find an expression for u_n when (i) $u_1 = 2$ and $u_{n+1} = nu_n - n^2 + 2$; (ii) $u_1 = 1$, $u_2 = 3$, and $u_{n+2} = 2u_{n+1} - u_n$.

18. Find a prime number which is a factor of all members of the sequence $2^{n+2} + 3^{2n+1}$.

19. Prove that any whole number greater than 15 can be expressed in the form $3a + 5b$, where a and b are positive integers.

20. If $u_{n+1} = 2(u_1 + u_2 + u_3 + \ldots + u_n)$ and $u_1 = 2$, prove that $u_2, u_3, \ldots u_n$ are the terms of a G.S. What is the value of $(u_2 + u_3 + u_4 + \ldots + u_n)$?

21. Prove that the number of ways of arranging n different articles in a row is $n!$

22. Prove that a set of n straight lines in a plane, no two of which are parallel and no three of which pass through one point, meet in $\frac{1}{2}n(n - 1)$ points.

23. Into how many parts is space divided by n planes all of which pass through the same point, but no three of which intersect in the same line?

24. $ABDC$ is a trapezium with $AB > CD$; AC, BD produced meet at O. P_n is on AB and $AP_n = (1/n)AB$. AD meets CP_n at X. OX produced meets AB at P_{n+1}. Calculate AP_{n+1} in terms of AB.
Hence design a construction, with a straight edge only, for dividing a line AB into n equal parts, given CD parallel to AB.

Summation of series

We have seen, in the previous section, how the method of mathematical induction can be used to check a proposed formula for the sum of n terms of a known sequence. Now we shall briefly consider a well-known method, the *method of differences*, which enables us to find the sum of n terms of many important sequences.

EXAMPLE. *Find* s_n, *where* $s_n = u_1 + u_2 + u_3 + \ldots + u_n$, *and*

$$u_p = \frac{1}{p(p + 1)}.$$

Since $\frac{1}{p} - \frac{1}{p+1} = \frac{1}{p(p+1)}$, we can write:

$$u_1 = \frac{1}{1.2} = 1 - \frac{1}{2},$$

$$u_2 = \frac{1}{2.3} = \frac{1}{2} - \frac{1}{3},$$

$$u_3 = \frac{1}{3.4} = \frac{1}{3} - \frac{1}{4},$$

. . . .

$$u_{n-1} = \frac{1}{(n-1)\,n} = \frac{1}{n-1} - \frac{1}{n},$$

$$u_n = \frac{1}{n\,(n+1)} = \frac{1}{n} - \frac{1}{n+1}.$$

Adding,

$$u_1 + u_2 + u_3 + \ldots + u_n = 1 - \frac{1}{n+1} = \frac{n}{n+1}.$$

There is an appeal to the inductive principle here, indicated by the row of dots, and, if desired, a formal proof by induction could now be given, $P(n)$ being the statement

$$s_n = \frac{n}{n+1}.$$

The sigma notation

It is customary to write $\sum_{r=1}^{r=n} u_r$ or $\sum_{r=1}^{n} u_r$ to indicate the sum

$$u_1 + u_2 + u_3 + \ldots + u_n.$$

Even more briefly, when there is no doubt about the letter which is to take up the successive values 1, 2, 3, . . . , n, we can write $\sum_1^n u_r$.

Ex. 6. Write out in full:

(i) $\sum_{r=1}^{4} \frac{1}{r}$; (ii) $\sum_{r=3}^{5} r(r+1)$; (iii) $\sum_{r=0}^{4} (-2)^r$.

Ex. 7. Use the sigma notation to abbreviate:

(i) $1^2 + 2^2 + 3^2 + 4^2$; (ii) $\frac{1}{2} - \frac{1}{3} + \frac{1}{4} - \frac{1}{5} + \frac{1}{6}$;

(iii) $1^3 - 2^3 + 3^3 - 4^3 + \ldots - (2n)^3$;

and show that $\sum_{r=1}^{3} \frac{1}{n+2} = \sum_{r=3}^{5} \frac{1}{n}$.

The difference method

Suppose, in the sequence $\{u_r\}$, we can express u_r in the form $v_r - v_{r-1}$, so that

$$u_1 = v_1 - v_0, \quad u_2 = v_2 - v_1, \quad u_3 = v_3 - v_2, \ldots,$$
$$u_{n-1} = v_{n-1} - v_{n-2}, \quad u_n = v_n - v_{n-1}.$$

Then $\sum_{r=1}^{n} u_r = v_n - v_0$, and a formula for $\sum_{1}^{n} u_r$ has been established. There is just one source of difficulty: we must ensure that the form of v_r allows the value $r = 0$. It is customary to overcome this by testing the expression for u_1.

The summation $\sum u_r$ where u_r has one of the forms r, $r(r+1)$, $r(r+1)(r+2)$, and so on, is important and easily dealt with by the difference method.

For example:

(i) $\sum_{r=1}^{n} r = \sum_{r=1}^{n} \{\frac{1}{2}r(r+1) - \frac{1}{2}(r-1)\,r\} = \frac{1}{2}n(n+1)$.

$u_1 = 1 = \frac{1}{2}.1.2 - \frac{1}{2}.0.1$, so that the difference relation $u_1 = v_1 - v_0$ is correct when $r = 1$.

(ii) $\sum_{r=1}^{n} r(r+1) = \sum_{r=1}^{n} \{\frac{1}{3}r(r+1)(r+2) - \frac{1}{3}(r-1)\,r\,(r+1)\}$

$= \frac{1}{3}n(n+1)(n+2)$.

$u_1 = 1.2 = \frac{1}{3}1.2.3 - \frac{1}{3}0.1.2$, and so $u_1 = v_1 - v_0$ is correct when $r = 1$.

Ex. 8. Find $\sum_{r=1}^{n} r(r+1)(r+2)$.

The allied series

$$\sum_{r=1}^{n} \frac{1}{r}, \quad \sum_{r=1}^{n} \frac{1}{r(r+1)}, \quad \sum_{r=1}^{n} \frac{1}{r(r+1)(r+2)}$$

are also important. There is no simple form for $\sum_{r=1}^{n} 1/r$, but difference formulae $u_r = v_r - v_{r+1}$ exist for the others.

(i) $$\sum_{r=1}^{n} \frac{1}{r(r+1)} = \sum_{r=1}^{n} \left\{\frac{1}{r} - \frac{1}{r+1}\right\} = 1 - \frac{1}{n+1} = \frac{n}{n+1}.$$

$u_1 = \dfrac{1}{1.2} = 1 - \frac{1}{2} = v_1 - v_2$, giving a correct form.

(ii) $$\sum_{r=1}^{n} \frac{1}{r(r+1)(r+2)} = \sum_{r=1}^{n} \frac{1}{2}\left\{\frac{1}{r(r+1)} - \frac{1}{(r+1)(r+2)}\right\}$$

$$= \frac{1}{4} - \frac{1}{2(n+1)(n+2)}.$$

$u_1 = \dfrac{1}{1.2.3} = \dfrac{1}{2}\cdot\dfrac{1}{1.2} - \dfrac{1}{2}\cdot\dfrac{1}{2.3} = v_1 - v_2$, again correct.

Ex. 9. Find $\sum_{r=1}^{n} \frac{1}{r(r+1)(r+2)(r+3)}$.

EXAMPLE. *Find* $\sum_{r=1}^{n} r^2$.

$$r^2 = r(r+1) - r \Rightarrow \sum_{r=1}^{n} r^2 = \sum_{r=1}^{n} r(r+1) - \sum_{r=1}^{n} r.$$

Hence

$$\sum_{r=1}^{n} r^2 = \tfrac{1}{3}n(n+1)(n+2) - \tfrac{1}{2}n(n+1)$$
$$= \tfrac{1}{6}n(n+1)\,\{(2n+4)-3\},$$

and so
$$\sum_{r=1}^{n} r^2 = \tfrac{1}{6}n(n+1)\,(2n+1).$$

EXAMPLE. *Find* $\displaystyle\sum_{r=1}^{n} \frac{1}{r(r+2)}$.

We have

$$\frac{1}{r(r+2)} = \frac{r+1}{r(r+1)(r+2)}$$
$$= \frac{(r+2)-1}{r(r+1)(r+2)}$$
$$= \frac{1}{r(r+1)} - \frac{1}{r(r+1)(r+2)}.$$

Hence

$$\sum_{r=1}^{n} \frac{1}{r(r+2)} = \sum_{r=1}^{n} \frac{1}{r(r+1)} - \sum_{r=1}^{n} \frac{1}{r(r+1)\,(r+2)}$$
$$= \left\{1 - \frac{1}{n+1}\right\} - \left\{\frac{1}{4} - \frac{1}{2(n+1)(n+2)}\right\}$$
$$= \frac{3}{4} - \frac{2(n+2)-1}{2(n+1)(n+2)},$$

and so

$$\sum_{r=1}^{n} \frac{1}{r(r+2)} = \frac{3}{4} - \frac{2n+3}{2(n+1)(n+2)}.$$

Ex. 10. Find $\sum_{r=1}^{n} r^3$ by writing $r^3 = r(r+1)(r+2) - 3r(r+1) + r$.

Ex. 11. Find $\sum_{r=1}^{n} \frac{1}{r(r+3)}$. Verify your result by induction.

The limit sum of a series

Suppose s_n represents the sum of the terms

$$u_1 + u_2 + u_3 + \ldots + u_n.$$

The behaviour of s_n, as more and more terms are taken, is frequently of considerable importance.

Suppose

$$s_n = \sum_{r=1}^{n} \frac{1}{r(r+1)} = 1 - \frac{1}{n+1}.$$

The sum s_n is clearly always less than 1, but

$$\lim_{n\to\infty} s_n = 1.$$

It is important to understand what is meant by $\lim_{n\to\infty} s_n$, which is often called the *limit sum* of the series. In no way is it to be considered as the actual sum of a series or as an actual value of s_n. If all the terms of s_n are positive, $\lim_{n\to\infty} s_n$ provides an *upper bound* to the value of s_n, and an upper bound which is never reached. If the terms of s_n are alternately positive and negative, $\lim_{n\to\infty} s_n$ gives a *point of concentration* about which the values of s_n cluster more and more closely, but never actually reach. These two cases are clearly exemplified by the series:

(i) $1 + \frac{1}{2} + \frac{1}{4} + \frac{1}{8} + \ldots + \frac{1}{2^{n-1}} + \ldots,$

for which

$$s_n = 2 - \frac{1}{2^{n-1}}, \quad \lim_{n\to\infty} s_n = 2;$$

$$\text{(ii) } 1-\frac{1}{2}+\frac{1}{4}-\frac{1}{8}+\ldots+\frac{(-1)^{n-1}}{2^{n-1}}+\ldots,$$

for which

$$s_n=\frac{2}{3}-\frac{2}{3}\left(-\frac{1}{2}\right)^n, \quad \lim_{n\to\infty} s_n=\frac{2}{3}.$$

Ex. 12. Draw graphs showing the values of s_n for $n = 1, 2, 3, 4, 5$ in the series (i) and (ii) above.

EXERCISE 10.2

In nos. 1–4 *the value of* u_r *is given, and* $s_n = \sum_{r=1}^{n} u_r$. *Find* s_1, s_2, s_3 *and* s_n.

1. $3r$. **2.** $r(r + 2)$. **3.** $2^r - 1$. **4.** $1/r(r + 2)$.

In nos. 5–8 *the value of* s_n *is given, where* $s_n = \sum_{r=1}^{n} u_r$. *Find* u_1, u_2, u_3 *and* u_n.

5. n^2. **6.** $3n^3 + 2n^2$. **7.** $3^n + n$. **8.** $\log (n + 1)$.

9. Use relations of the type $n(2n + 1) = 2n(n + 1) - n$ to find the sum of n terms of the sequences whose rth terms are:

(i) $r(r + 2)$; (ii) $2r^3 + r^2$; (iii) $(2r - 1)(2r + 3)$.

10. Use the identities $n^2 - (n - 1)^2 = 2n - 1$, $n^3 - (n - 1)^3 = 3n^2 - 3n + 1$ and the difference method to find

$$\sum_{r=1}^{n} r \quad \text{and} \quad \sum_{r=1}^{n} r^2.$$

11. Use the identity $n(n + 2)(n + 4) - (n - 2)n(n + 2) = 6n(n + 2)$ to find $\sum_{r=1}^{n} (2r - 1)(2r + 1)$ and $\sum_{r=1}^{n} 2r(2r + 2)$.

12. Use $\dfrac{2}{(2n-1)(2n+1)} = \dfrac{1}{2n-1} - \dfrac{1}{2n+1}$ to find $\displaystyle\sum_{r=1}^{n} \frac{1}{4r^2-1}$.

In nos. 13–18 *an expression is given for* u_r; *find* $\sum_{r=1}^{n} u_r$ *in each case:*

13. $(2r + 1)(2r + 3)$. **14.** $r(2r + 3)$. **15.** $3r^2 - 5r$.

16. $\dfrac{1}{(3r-2)(3r+1)}$. **17.** $\dfrac{1}{(2r-1)(2r+3)}$. **18.** $\log\dfrac{r+2}{r}$.

19. Find the limit sum of the following:

(i) $\displaystyle\sum_{r=1}^{n}\frac{1}{(3r-1)(3r+2)}$; (ii) $\displaystyle\sum_{r=1}^{n}\frac{r}{(r+1)!}$; (iii) $\displaystyle\sum_{r=1}^{n}\frac{r+3}{r(r+1)(r+2)}$.

20. Show that $\sum_{r=1}^{n} r(r+1)(r+2)$ is a quartic polynomial in n which is always divisible by 6. Generalise this statement.

CHAPTER 11

DIFFERENTIATION

Gradient of a chord

Suppose a car is travelling along a road and recordings are made of two variables:

(i) the time at which the observation is made, measured from a certain instant such as the instant at which a stop-watch was started;

(ii) the distance of the car from a fixed point on the road such as a milestone.

The time (t seconds) will be taken as the independent variable, the distance (s feet) as the dependent variable. Figure 11.1 represents the graph of s regarded as a function of t.

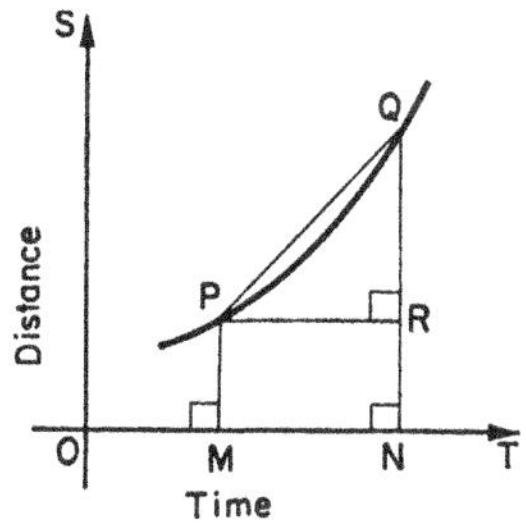

Fig. 11.1

In Fig. 11.1, M and N correspond to two values of t, t_1 and t_2. PM and QN represent s_1 and s_2, the corresponding values of s. The average velocity of the car in the interval of time MN is

$$\frac{s_2 - s_1}{t_2 - t_1} \text{ ft/sec.}$$

This fraction also gives the gradient of the chord PQ in the graph provided that equal lengths on the two axes represent unit difference of time and unit difference of distance.

In a distance–time graph we therefore associate average speed over a certain time interval with the gradient of the corresponding chord in the graph.

Ex. 1. Find the average speed over the given time-intervals when s and t are connected by the relation $s = t^2$:

(i) $2 \leqslant t \leqslant 3, \quad 2 \leqslant t \leqslant 2{\cdot}1, \quad 2 \leqslant t \leqslant 2 + h;$

(ii) $1 \leqslant t \leqslant 2, \quad 1{\cdot}9 \leqslant t \leqslant 2, \quad 2 - h \leqslant t \leqslant 2.$

Gradient of a tangent

Figure 11.2 also represents a portion of a distance–time graph. The gradient of the chord Q_1P gives the average velocity of the car over the time-interval N_1M, and the gradient of the chord PQ_2 gives the average velocity over the time-interval MN_2.

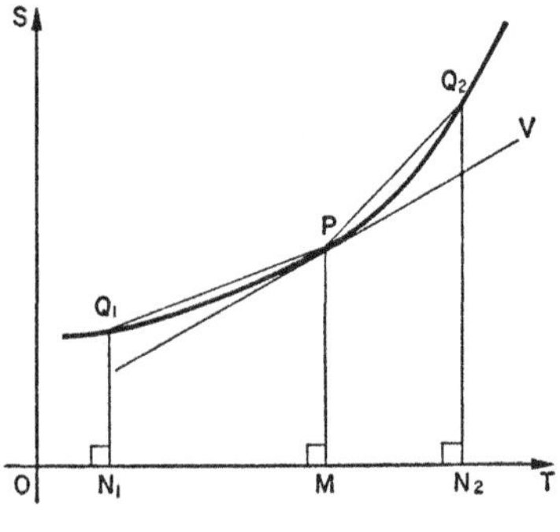

Fig. 11.2

But there is a line PV passing through P which does not meet the curve again in the neighbourhood of P; it is the tangent at P. Its gradient is greater than the gradient of any chord Q_1P drawn from a point Q_1 to the left of P, and is less than the gradient of any chord drawn to a point Q_2 to the right of P. We say that the gradient of PV is *the gradient of the curve* at the point P, and it gives the *velocity at the instant* represented by the point M.

As the time represented by MN_2 gets smaller the gradient of the chord PQ_2 gets nearer to the value of the gradient of the tangent at P.

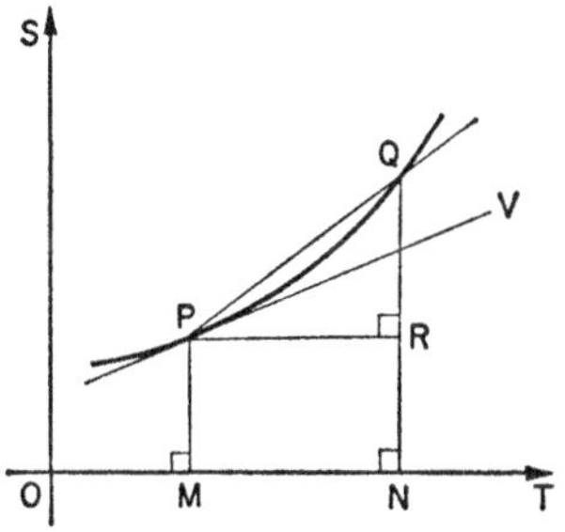

Fig. 11.3

In symbols we say that the gradient of the tangent PV (see Fig. 11.3) is

$$\lim_{PR \to 0} \left(\frac{QR}{PR}\right).$$

Gradient of a point on the curve $y = x^2$

Figure 11.4 represents a portion of the graph of $y = x^2$ (not

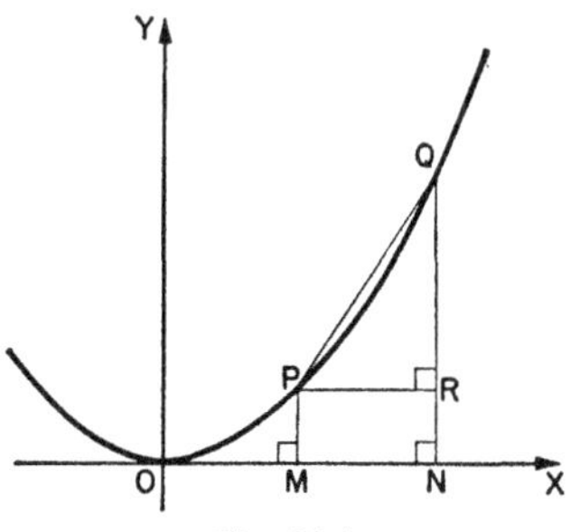

Fig. 11.4

drawn to scale). Suppose P is the point (2, 4) and let $MN = h$; then $QN = (2 + h)^2$. Hence

$$\begin{aligned} QR &= QN - PM \\ &= (2 + h)^2 - 4 \\ &= 4h + h^2, \end{aligned}$$

and

$$\frac{QR}{PR} = \frac{4h + h^2}{h} = 4 + h.$$

Similarly, if Q were the point for which $x = 2 - h$, we should find

$$\frac{QR}{PR} = 4 - h.$$

All chords drawn from P must therefore have a gradient greater than 4, or a gradient less than 4. By taking h small enough, we can make both $(4 + h)$ and $(4 - h)$ as near to 4 as we please. The only line through P with the gradient 4 exactly is the tangent at P, and so the gradient of $y = x^2$ at the point (2, 4) is 4.

EXERCISE 11.1

1. Find the gradient of the chord AB of the curve $y = x^2$ when the coordinates of A and B are as follows:

(i) A (1, 1), B (2, 4); (ii) A (−2, 4), B (1, 1); (iii) A (0, 0), B (1·5, 2·25); (iv) A (−3, 9), B (−2, 4).

2. A is the point (1, 1) on the curve $y = x^2$. If the x-coordinate of B is $1 + h$, find the gradient of AB when h is:

(i) 1; (ii) 0·5; (iii) 0·2; (iv) 0·1; (v) 0·01; (vi) 0·001.

3. A car entered a motorway, and t seconds after entering it had travelled s feet from the point of entry. During the first 5 sec the relation between s and t was $s = 30t + 2t^2 + \frac{1}{3}t^3$. Find the average speed of the car over the following time-intervals:

(i) $0 \leqslant t \leqslant 1$; (ii) $0 \leqslant t \leqslant 3$; (iii) $1 \leqslant t \leqslant 3$; (iv) $1 \leqslant t \leqslant 1{\cdot}1$.

4. Find the gradient of the chord AP of the curve $y = 1/x$ when A is (1, 1) and the x-coordinate of P is $1 + h$, where h is:

(i) 2; (ii) 1; (iii) 0·5; (iv) 0·1; (v) 0·01.

Gradient formula

We can calculate the gradient at the point (x, y) on the graph of $y = x^2$ in the same way as we calculated the gradient at (2, 4).

The symbol Δx is used to denote a small change in the value of x. In Fig. 11.5, $OM = x$ and $MN = \Delta x$. The symbol Δy is used to denote the change in the value of y corresponding to the change in x from x to $x + \Delta x$. In Fig. 11.5, $PM = y$ and $QN = y + \Delta y$, so that $QR = \Delta y$. Δx is to be treated as a single symbol and it is often called the *increment* of x; Δy is the corresponding increment of y. An increment can be either positive or negative.

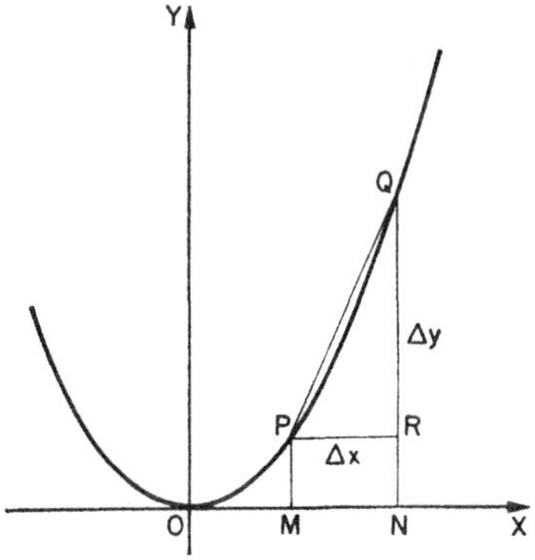

Fig. 11.5

Figure 11.5 again represents the graph of $y = x^2$. P is (x, y) and Q is $(x + \Delta x, y + \Delta y)$. But $PM = OM^2$ and $QN = ON^2$, so that

$$\begin{aligned}\Delta y = QR &= QN - PM \\ &= (x + \Delta x)^2 - x^2 \\ &= 2x(\Delta x) + (\Delta x)^2,\end{aligned}$$

$$\frac{\Delta y}{\Delta x} = 2x + \Delta x.$$

In Fig. 11.6, N_1N_2 represents the Δx-neighbourhood of M, where $OM = x$.

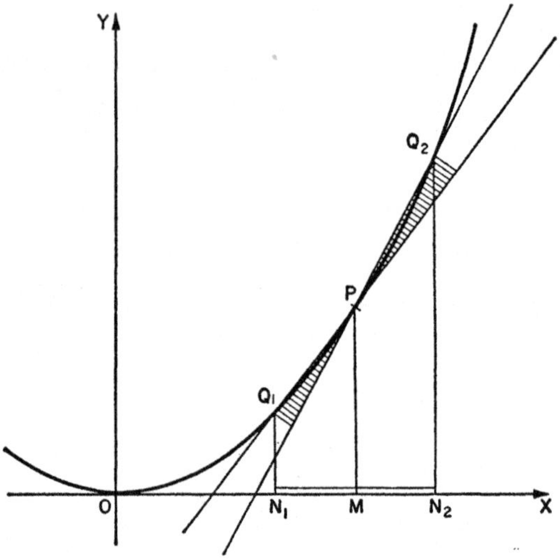

Fig. 11.6

The chords PQ_1 and PQ_2 then define a corresponding neighbourhood of P in the plane XOY, indicated by the shaded area. The line through P with gradient $2x$ lies within this neighbourhood for all small values of Δx, and we write

$$\lim_{\Delta x \to 0} \left(\frac{\Delta y}{\Delta x}\right) = 2x.$$

Hence the gradient of the curve $y = x^2$ at the point (x, y) on the curve is given by $2x$. For the present we shall denote this gradient by $\mathrm{D}y$, so that, for $y = x^2$:

$$\mathrm{D}y = \lim_{\Delta x \to 0} \left(\frac{\Delta y}{\Delta x}\right) = 2x.$$

A gradient-formula found in this way is known as the *derived function*; it is also called the *derivative*. The derived function of x^2 is written $\mathrm{D}(x^2)$.

Similarly, for the general function $y = f(x)$, the derived function is written Dy, or D$\{f(x)\}$, or $f'(x)$. The process of finding the derived function is called *differentiation*.

Ex. 2. Use the above method to differentiate (i) $2x^2$, (ii) $3x + 2$, (iii) $x^2 + 5x$.

Differentiation of powers of x

(i) Let $y = c$, a constant.
The graph of $y = c$ is a line parallel to Ox, as in Fig. 11.7.

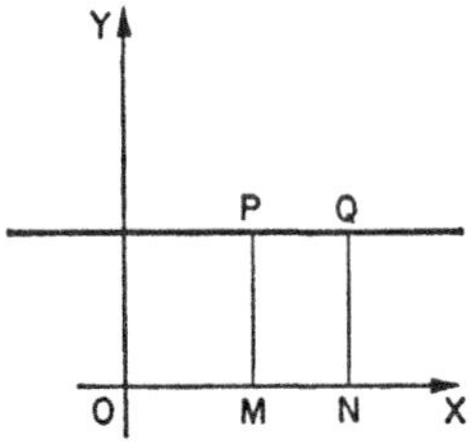

Fig. 11.7

Then, if P is (x, y) and $MN = \Delta x$,

$$\frac{\Delta y}{\Delta x} = \frac{0}{\Delta x}$$

for all values of Δx. Hence Dy = 0, since the neighbourhood of P in Fig. 11.7 is reduced to a single line.

(ii) Let $y = bx$, where b is constant.
The graph is a straight line through O, as in Fig. 11.8. Then, if

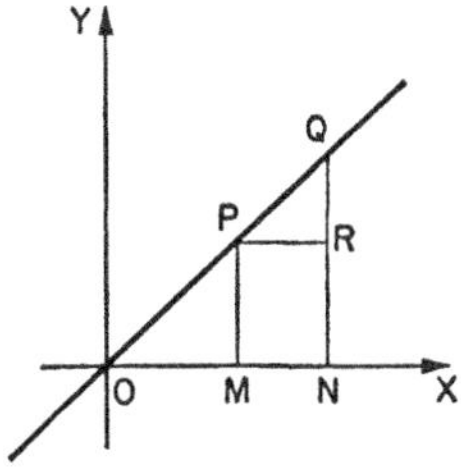

Fig. 11.8

P is (x, y) and $MN = \Delta x$, $RQ = \Delta y$,

$$\frac{\Delta y}{\Delta x} = \frac{QN - PM}{\Delta x} = \frac{b(x + \Delta x) - bx}{\Delta x} = b$$

for all values of Δx. Hence $\mathrm{D}y = b$.

(iii) Let $y = ax^2$, where a is constant.
The graph is a parabola, as in Fig. 11.9.

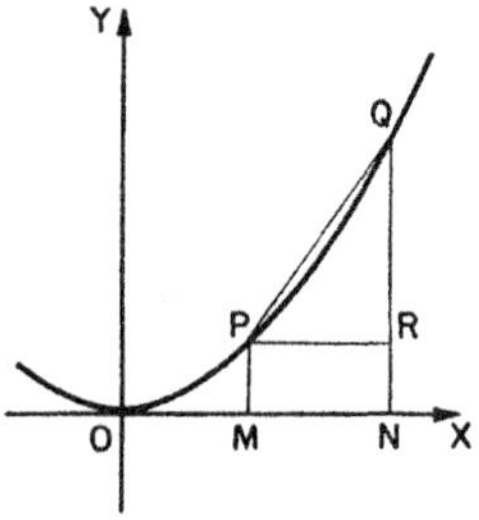

Fig. 11.9

$$\frac{\Delta y}{\Delta x} = \frac{QN - PM}{\Delta x}$$

$$= \frac{a(x + \Delta x)^2 - ax^2}{\Delta x}$$

$$= \frac{2ax(\Delta x) + a(\Delta x)^2}{\Delta x}$$

$$= 2ax + a(\Delta x).$$

Hence

$$\mathrm{D}y = \lim_{\Delta x \to 0} \left(\frac{\Delta y}{\Delta x}\right) = 2ax.$$

(iv) Let

$$y = ax^2 + bx + c,$$

$$y + \Delta y = a(x + \Delta x)^2 + b(x + \Delta x) + c,$$

$$\Delta y = 2ax(\Delta x) + (\Delta x)^2 + b(\Delta x),$$

$$\frac{\Delta y}{\Delta x} = 2ax + \Delta x + b.$$

$$Dy = \lim_{\Delta x \to 0} \left(\frac{\Delta y}{\Delta x}\right) = 2ax + b.$$

It is important to note that this result is the same as if we had differentiated the three terms ax^2, bx, c separately. It is a special case of the important principle:

If $y = f(x) + g(x)$, then $Dy = f'(x) + g'(x)$.

(v) Let

$$y = \frac{k}{x}, \quad \text{where } k \text{ is constant,}$$

$$y + \Delta y = \frac{k}{x + \Delta x},$$

$$\Delta y = \frac{k}{x + \Delta x} - \frac{k}{x}$$

$$= \frac{kx - k(x + \Delta x)}{x(x + \Delta x)}$$

$$= \frac{-k(\Delta x)}{x(x + \Delta x)},$$

$$\frac{\Delta y}{\Delta x} = -\frac{k}{x(x + \Delta x)}.$$

$$Dy = \lim_{\Delta x \to 0} \left(\frac{\Delta y}{\Delta x}\right) = -\frac{k}{x^2}.$$

Another general principle, which governs (i) to (v) above, is:

If $y = kf(x)$, where k is constant, $Dy = kf'(x)$.

EXAMPLE. *Differentiate the function* $\frac{1}{2x + 3}$ *from first principles (that is, using the* Δ*-process).*

Let

$$y = \frac{1}{2x + 3},$$

$$y + \Delta y = \frac{1}{2(x + \Delta x) + 3}.$$

$$\Delta y = \frac{1}{2(x + \Delta x) + 3} - \frac{1}{2x + 3}$$

$$= \frac{(2x + 3) - 2(x + \Delta x) - 3}{(2x+3)\ \{2(x + \Delta x) + 3\}}$$

$$= - \frac{2(\Delta x)}{(2x + 3)\ \{2(x + \Delta x) + 3\}},$$

$$\frac{\Delta y}{\Delta x} = - \frac{2}{(2x + 3)\ \{2(x + \Delta x) + 3\}}.$$

$$\mathrm{D}y = \lim_{\Delta x \to 0} \left(\frac{\Delta y}{\Delta x}\right) = - \frac{2}{(2x + 3)^2}.$$

EXAMPLE. *A particle travelling along a straight path covers a distance s metres in t seconds, where* $s = 1500t + 400t^2$. *Find its speed after* 2 *seconds.*

$$s = 1500t + 400t^2,$$

$$\mathrm{D}s = 1500 + 800t.$$

After 2 seconds, $\mathrm{D}s = 1500 + 1600 = 3100$. At the end of 2 seconds the speed of the particle is 3100 metres per second.

EXERCISE 11.2

In nos. 1–6 *differentiate the given function from first principles, using the* Δ*-process:*

1. $3x^2 + 2$. **2.** x^3. **3.** $\dfrac{4}{3 - 2x}$. **4.** x^4. **5.** $\dfrac{1}{x^2}$. **6.** $\dfrac{2}{(x + 1)^2}$.

7. Write down the derivatives of:

$$x^4,\ \ x^3,\ \ x^2,\ \ x,\ \ 1,\ \ 1/x,\ \ 1/x^2.$$

Does there appear to be a formula for the derivative of x^n which applies in all these seven cases?

8. Find the gradient of the curve $y = x^2 - 3x + 4$ at the point where $x = 2$. Write down the equation of the tangent to the curve at this point.

9. Find the equation of the tangents to the following curves at the given points:

(i) $y = 6 - x + x^2$ where $x = 3$; (ii) $y = 2/x$ where $x = 4$.

10. Find the point on the curve $y = 3x^2 - 2x - 5$ where the gradient is 10, and the point where the gradient is zero.

11. Find the points on the given curves with the required gradients:

(i) $y = 4x^3 - 5x$, gradient -2;

(ii) $y = \frac{1}{2x^2} - \frac{1}{x}$, gradient -4.

12. A particle has travelled a distance s metres in time t seconds. Find its speed when

(i) $s = 20 - t + 3t^2, t = 3$; (ii) $s = 90t - 16t^2, t = 1{\cdot}5$.

The derivative of x^n

We have seen in no. 7 of Exercise 11.2, on p. 164, that, when n has any integral value from -2 to $+4$,

$$D(x^n) = nx^{n-1}.$$

This rule is true for all real values of n, but the general proof will not be attempted here. We shall verify it in a few particular cases.

EXAMPLE. *Differentiate* $\sqrt{x}$.

$$y = \sqrt{x},$$

$$y + \Delta y = \sqrt{(x + \Delta x)},$$

$$\Delta y = \sqrt{(x + \Delta x)} - \sqrt{x}$$

$$= \frac{(x + \Delta x) - x}{\sqrt{(x + \Delta x)} + \sqrt{x}},$$

$$\frac{\Delta y}{\Delta x} = \frac{1}{\sqrt{(x + \Delta x)} + \sqrt{x}}.$$

Hence

$$Dy = \lim_{\Delta x \to 0} \left\{\frac{1}{\sqrt{(x + \Delta x)} + \sqrt{x}}\right\} = \frac{1}{2\sqrt{x}}$$

We note that $D(x^{1/2}) = \frac{1}{2}x^{-1/2}$ follows the general rule

$$D(x^n) = nx^{n-1}.$$

Ex. 3. Use the general rule to write down the derived functions of (i) x^8; (ii) $\sqrt{x^3}$; (iii) $1/\sqrt{x}$; (iv) $\sqrt[3]{x}$; (v) $1/x^5$.

The chain rule for differentiation

If we wish to differentiate an expression like $(2x^2+3)^3$ we can multiply out and then differentiate term by term. This is tedious, and we are naturally led to consider the substitution $u = 2x^2+3$.

Suppose $y = (2x^2+3)^3 = u^3$. The dependent variable y is now expressed as a function of u; the derivative of y with respect to u is written $D_u y$ and the derivative of y with respect to x is written $D_x y$. Also u is a function of x; the derivative of u with respect to x is written $D_x u$.

Consider the effect of an increment Δx. It produces an increment $\Delta u = 4x(\Delta x)+2(\Delta x)^2$, which in turn produces an increment $\Delta y = 3u^2(\Delta u)+3u(\Delta u)^2+(\Delta u)^3$. Clearly,

$$\Delta x \to 0 \Rightarrow \Delta u \to 0 \Rightarrow \Delta y \to 0.$$

Then

$$\frac{\Delta y}{\Delta u} - D_u y = \epsilon, \text{ where } \epsilon \to 0 \text{ as } \Delta x \to 0,$$

$$\Delta y = (D_u y)\,\Delta u + \epsilon(\Delta u),$$

$$\frac{\Delta y}{\Delta x} = (D_u y)\,.\,\frac{\Delta u}{\Delta x} + \epsilon\,.\,\frac{\Delta u}{\Delta x}.$$

Hence

$$D_x y = \lim_{\Delta x \to 0}\left(\frac{\Delta y}{\Delta x}\right)$$

$$= D_u y\,.\,D_x u + 0\,.\,D_x u$$

and so

$$\mathbf{D_x y = D_u y\,.\,{}_x D u.}$$

Since

$$y = u^3 \text{ and } u = 2x^2+3,$$

$$D_x y = 3u^2(4x) = 12x(2x^2+3)^2.$$

If z were now a function of y, say $(y + 1)^2$, we could extend the argument to show that

$$D_x z = D_y z \, . \, D_u y \, . \, D_x u.$$

Thus, if $z = \{(2x^2 + 3)^3 + 1\}^2$,

$D_x z = (2y + 2) \, . \, 3u^2 \, . \, 4x = 24x(2x^2 + 3)^2 \, \{(2x^2 + 3)^3 + 1\}$.

It will now be clear why $D_x y = D_u y \, . \, D_x u$ is called the *chain rule for differentiation.*

A general proof of the chain rule when $y = f(u)$ and $u = g(x)$ is beyond the scope of this book since it involves more advanced theorems on limits than we have dealt with in Chapter 9, but the reader can apply the rule with confidence to all the functions which he will be expected to differentiate at this stage.

Implicit differentiation

The Δx process of differentiation can be applied to the individual terms of a relation which implicitly defines y as a function of x. For example, if $x^2 + y^2 = 4$, then

$$D_x(x^2) + D_x(y^2) = D_x(4). \tag{i}$$

If $z = y^2$, $D_y z = 2y$ so $D_x z = 2y(D_x y)$, from the chain rule. Hence (i) gives

$$2x + 2y(D_x y) = 0,$$

or

$$D_x y = -\frac{x}{y}.$$

EXAMPLE. *Differentiate* $\surd(2x^2 + 3)$.

Let $y = (2x^2 + 3)^{1/2}$ and $u = 2x^2 + 3$. Then

$$y = u^{1/2}, \quad D_u y = \tfrac{1}{2}u^{-1/2} = \frac{1}{2\sqrt{u}}.$$

Hence

$$D_x y = D_u y \, . \, D_x u$$

$$= \frac{1}{2\sqrt{u}} \, . \, 4x,$$

giving

$$D_x\{\sqrt{(2x^2+3)}\} = \frac{2x}{\sqrt{(2x^2+3)}}.$$

Ex. 4. Check the above example by implicit differentiation of $y^2 = 2x^2 + 3$.

Ex. 5. Use the chain rule to differentiate (i) $(x + 2)^4$, (ii) $(x + 1/x)^3$, (iii) $\sqrt{(x^3 + 1)}$, (iv) $\sqrt{(2x - 3)^3}$.

EXAMPLE. *Find the equation of the tangent at the point* (4, 9) *to the curve* $x^{1/2} + y^{1/2} = 5$.

Implicit differentiation of $x^{1/2} + y^{1/2} = 5$ gives

$$\tfrac{1}{2}x^{-1/2} + \tfrac{1}{2}y^{-1/2}(D_x y) = 0.$$

If m is the value of $D_x y$ at (4, 9),

$$\tfrac{1}{2}(\tfrac{1}{2}) + \tfrac{1}{2}(\tfrac{1}{3}) \, . \, m = 0,$$

and so $m = -\frac{3}{2}$.

Hence the tangent to the given curve at (4, 9) has a gradient $-\frac{3}{2}$. Its equation is therefore:

$$y - 9 = -\tfrac{3}{2}(x - 4)$$

or

$$3x + 2y = 30.$$

The chain rule allows us to verify further examples of the general rule $D(x^n) = nx^{n-1}$.

Suppose $y = x^{5/2}$, and let $u = x^{1/2}$. Then

$$y = u^5 \text{ and } D_u y = 5u^4.$$

But

$$D_x x^{1/2} = \tfrac{1}{2}x^{-1/2},$$

and so

$$D_x y = D_u y \, . \, D_x u$$

$$= 5u^4 \, . \, \tfrac{1}{2}x^{-1/2},$$

or

$$D_x(x^{5/2}) = \tfrac{5}{2}x^{3/2},$$

which follows the general rule.

Ex. 6. Verify that $D_x(x^{-5}) = -5x^{-6}$, assuming that $D_x(1/x) = -x^{-2}$.

EXERCISE 11.3

1. Write down the derivatives of the following functions:
(i) $3x^3 + 2$; (ii) $4x - 5x^2$; (iii) $3 + 5x^2 - 2x^4$; (iv) $3 - 1/x$;

(v) $\dfrac{2}{x} - \dfrac{3}{x^2}$; (vi) $\dfrac{4x-5}{x^2}$; (vii) $4\sqrt{x} - 5$; (viii) $3 - \dfrac{2}{x\sqrt{x}}$;

(ix) $\dfrac{2-\sqrt{x}}{x}$.

2. Use a substitution and the chain rule to differentiate:

(i) $(2x + 3)^4$; (ii) $\sqrt{(3x - 4)}$; (iii) $\dfrac{1}{5-3x}$; (iv) $(3x^2 + 2)^3$;

(v) $\sqrt{(x^2 - 5)}$; (vi) $\dfrac{1}{3-2x^2}$; (vii) $\dfrac{3x-1}{\sqrt{(3x-2)}}$; (viii) $\dfrac{x^2}{(x^2+2)^{3/2}}$;

(ix) $\dfrac{2(1-x^2)}{(3-2x^2)^2}$.

3. Find the gradients of the curves at the given points:
(i) (2, 1), $2x^2 + 3y^2 = 11$; (ii) (3, 2), $3x^2 - 2y^2 = 19$;
(iii) (2, 3), $2/x + 3/y = 2$; (iv) (8, 1), $x^{2/3} + y^{2/3} = 5$.

4. Find the equations of the tangents to the curves at the given points:
(i) (2, 1), $x^2y = 4$; (ii) $(a, 2a)$, $x^3 + y^3 = 9a^3$;
(iii) (2, −3), $4/x - 3/y = 3$; (iv) (−2, −1), $xy = 2x + 6$.

5. Find the equations of the tangents to the curves
(i) $y^2 = 4x$ which are parallel to $x - y = 2$;
(ii) $x^2 + y^2 = 25$ which are perpendicular to $3x - 4y = 10$.

6. A ladder, of length 25 ft, is placed with its lower end A on horizontal ground and its upper end B in contact with a vertical wall. The end A slides directly away from the wall with a constant speed of 2 ft/sec. Find the rate of descent of B at the instant when A is 7 ft from the wall.

7. Find the equation of the tangent at the point P $(a, 1/a)$ to the curve $xy = 1$. If this tangent cuts the axis at A, B prove that P is the mid-point of AB.

CHAPTER 12

APPLICATIONS OF DIFFERENTIATION AND THE INVERSE PROCESS

Rate of change

If a car is driven along a road and its distance from a fixed point on the road after t seconds is s feet, we know that the speed at any instant is measured by the value of $D_t s$ for the corresponding value of t.

Speed is only one example of the concept of a *rate of change*, by which we normally mean time-rate. If a variable z is a function of the time t, the rate of change of z is the value of $D_t z$ at the moment under consideration.

On this analogy we sometimes speak of $D_x y$ as the rate of change of y with respect to x. We know that $D_t y = D_x y \,.\, D_t x$, and so

$$D_x y = D_t y / D_t x.$$

EXAMPLE. *A right circular cone has a slant height of* 13 *cm. The radius of the base at time t sec is x cm and the height is y cm. If x increases at the rate of* 2 *cm/sec, find the rate of increase of y when x is* 5.

Figure 12.1 shows a section of the cone through its axis at time t seconds. Then

$$x^2 + y^2 = 169,$$

$$2x + 2y\,(D_x y) = 0,$$

$$D_x y = -x/y.$$

When $x = 5$, $y = 12$ and $D_x y = -\frac{5}{12}$.

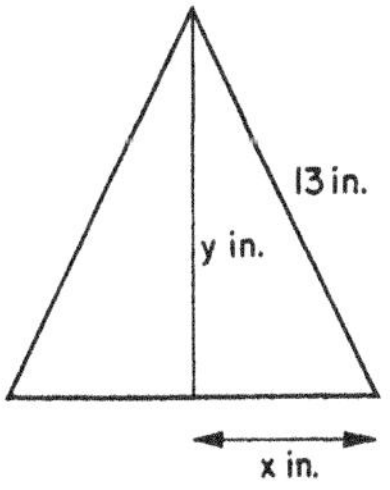

Fig. 12.1

Also $D_t x = 2$, so that, when $x = 5$,

$$D_t y = D_x y \,.\, D_t x$$

$$= -\frac{5}{12} \,.\, 2 = -\frac{5}{6}.$$

Hence y is decreasing at 5/6 cm/sec.

EXERCISE 12.1

1. Find the rate of change of A when $r = 5$ if $A = \pi r^2$ and $D_t r = 0{\cdot}5$.

2. Find the rate of increase of the surface (S cm^2) and the volume (V cm^3) of a sphere of radius r cm when $r = 3$ and r is increasing at 0·25 cm/sec. [$S = 4\pi r^2$, $A = \frac{4}{3}\pi r^3$.]

3. If $1/v + 1/u = 1/f$, where f is constant, find the rate of change of v when $u = 2$, $v = 5$ and $D_t u = 4$.

4. If $pv = 100$, find the rate of change of p when $v = 25$ and v is increasing at the rate of 0·5 units per second.

5. A sphere is expanding so that its surface area is increasing at the rate of 2 cm^2/sec. Find the rate of increase of (i) its radius, (ii) its volume, when the radius is 5 cm.

6. A container is designed to have a volume $(30x - x^3)$ cm^3 when the depth of water in it is x cm. Water is poured into the container at a steady rate of 5 cm^3/sec. At what rate is the water-level rising when the depth is 2 cm?
What is the area of the surface of the water when the depth is 2 cm. [Show that $D_x V = A$, where V cm^3 is the volume of water in the container and A cm^2 is the surface area.]

In nos. 7–12 find the expression for $D_x y$ *in terms of* t.

7. $x = t^2$, $y = 2t$. **8.** $x = t^2 - 1$, $y = t^3$. **9.** $x = ct$, $y = c/t$.

10. $x = a(1 - t^2), y = a(2 - t)(1 - t).$

11. $x = \sqrt{(1 + t)}, y = 2 + t^2.$

12. $x = \dfrac{1}{\sqrt{(1 - t^2)}}, y = t + 1/t.$

Stationary values of a function

If Fig. 12.2 represents a portion of the graph of $y = f(x)$, the value of Dy is zero when x has any one of the values x_1, x_2, x_3, x_5 corresponding to the points A, B, C, E. Remembering the use

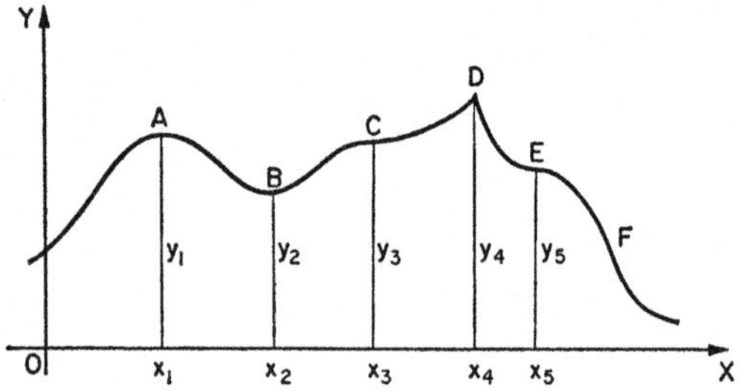

Fig. 12.2

of the derivative as a rate-measurer, the corresponding values of y (y_1, y_2, y_3, y_5) are called *stationary values* of the function $y = f(x)$, and the points A, B, C, E are called *stationary points.*

If $y = f(x)$, the values of x corresponding to the stationary points on the graph are found by solving the equation $f'(x) = 0$. In Fig. 12.2 the value $x = x_4$ will not be a root of $f'(x) = 0$; in fact we could not find $D_x y$ for $x = x_4$ because $\lim\limits_{\Delta x \to 0} (\Delta y/\Delta x)$ does not have a determinate value. We say that $f(x)$, although continuous at $x = x_4$, is not *differentiable.*

Points like A and B on a graph are important because they give *relative maximum* and *relative minimum* values of the function. Points like C, E, and F are called *points of inflection.* The maximum and minimum values of a function are naturally of great practical importance.

(1) *At a relative maximum the derived function vanishes and changes sign from positive to negative.*

(2) *At a relative minimum the derived function vanishes and changes sign from negative to positive.*

When $f'(x) > 0$ we say that $f(x)$ is an *increasing function*; the graph is rising as we move from left to right.

When $f'(x) < 0$ we say that $f(x)$ is a *decreasing function*; the graph is falling as we move from left to right.

The statement (1) above may be expressed more rigorously as follows:

> The function $y = f(x)$ is said to have a relative maximum at $x = x_1$ if and only if there exists an interval I surrounding x_1 such that $x \in I \Rightarrow f(x) \leqslant f(x_1)$, and $f(x) = f(x_1)$ only when $x = x_1$.

$f(x_1)$ is said to be an absolute maximum of $f(x)$ only if $f(x_1) \geqslant f(x)$ for all x in the domain of the function.

EXAMPLE. *Examine the stationary values of* $\dfrac{x}{1+x^2}$.

Let

$$y = \frac{x}{1+x^2} = \frac{1}{x+1/x}.$$

Put

$$u = x + \frac{1}{x}, \quad D_x u = 1 - \frac{1}{x^2}.$$

$$y = \frac{1}{u}, \quad D_u y = -\frac{1}{u^2}.$$

Hence

$$D_x y = D_x u \, . \, D_u y,$$

$$D_x y = -\frac{1 - 1/x^2}{(x+1/x)^2} = -\frac{(x+1)(x-1)}{(x^2+1)^2}.$$

But $-(x+1)(x-1)$ is positive when $-1 < x < 1$ and negative elsewhere.

Hence, when $x = -1$, $D_x y = 0$ and changes from $-$ to $+$ (see Fig. 12.3); y has a minimum value.

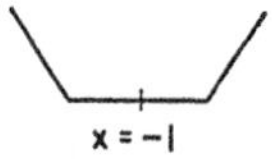

Fig. 12.3

When $x = 1$, $D_x y = 0$ and changes from + to − (see Fig. 12.4); y has a maximum value.

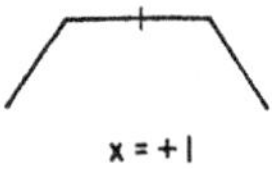

Fig. 12.4

The stationary values of y occur as follows:

(i) $x = -1$, $y = -\frac{1}{2}$, a relative minimum;
(ii) $x = +1$, $y = +\frac{1}{2}$, a relative maximum.

Ex. 1. Examine the stationary values of (i) $x^3 - 3x + 1$, (ii) $1/x + 1/x^2$. Sketch the graphs of the functions.

EXAMPLE. *Show that the rectangle of greatest area that can be inscribed in a circle is a square.*

Figure 12.5 shows a rectangle $ABCD$ inscribed in a circle with centre O. AC and BD must be diameters, since

$$A\hat{B}C = A\hat{D}C = 90°.$$

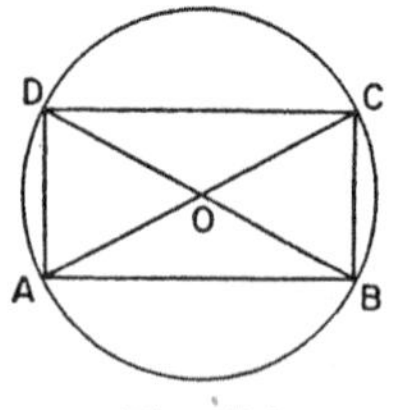

Fig. 12.5

Let $AB = x$ cm, $BC = y$ cm and $AC = d$ cm, where d is constant. Suppose the area of $ABCD$ is z cm^2, so that $z = xy$.

But $x^2 + y^2 = d^2$, so

$$z = x\sqrt{(d^2 - x^2)} = (d^2x^2 - x^4)^{1/2}.$$

Using the chain rule,

$$D_xz = \tfrac{1}{2}(d^2x^2 - x^4)^{-1/2} . (2d^2x - 4x^3),$$

$$D_xz = \frac{(d^2 - 2x^2)}{\sqrt{(d^2 - x^2)}}.$$

Since $x > 0$, the only stationary value of z occurs when

$$x = d/\sqrt{2}.$$

When $x = d/\sqrt{2}$, $D_xz = 0$ and changes from + to −. The absolute maximum of z therefore occurs when $x = d/\sqrt{2}$, and in this case $y = d/\sqrt{2}$ also.

Hence the rectangle of greatest area that can be inscribed in a given circle is a square.

EXERCISE 12.2

1. Find the values of x for which $y = x^2 - 4x + 5$ is (i) increasing, (ii) decreasing as x increases.

Find the stationary value of the function and state whether it is a maximum or minimum. Sketch the graph.

2. Find the intervals in which x must lie when $y = 2x^3 - 9x^2 + 12x - 4$ is (i) an increasing, (ii) a decreasing function.

Find the stationary values of the function and state their nature. Sketch the graph.

3. Find the stationary values of the following functions. In each case state the value of x, the value of the function and the nature of the stationary value.

(i) $2 - 3x^2 + 12x$; (ii) $\frac{1}{2}x^2 - 4x - 5$; (iii) $(x + 1)(x - 3)$;

(iv) $\sqrt{(x^2 + 2)} - \frac{1}{2}x^2$; (v) $\dfrac{x^2 + 1}{\sqrt{(x^2 + 2)}}$; (vi) $x + 1/x$.

4. Divide a line of length 10 cm into two parts of lengths x cm and y cm so that

(i) xy is greatest; (ii) $x^2 + y^2$ is greatest; (iii) $x^2 + 4y^2$ is least.

5. The perimeter of a rectangle is fixed. What is the shape of the rectangle when the length of the diagonal is least?

6. A piece of wire 10 cm long is bent to form three sides of a rectangle. Find the greatest possible area of the rectangle.

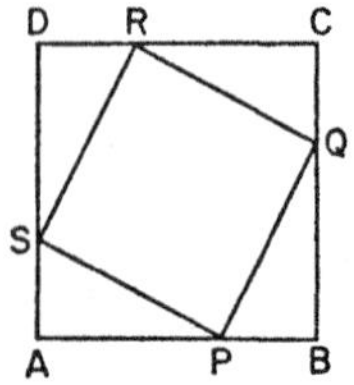

Fig. 12.6

7. A square *PQRS* is inscribed in a given square *ABCD* of side 5 cm, as in Fig. 12.6. Find the least possible area of *PQRS*.

8. A piece of wire in the form of a circle of radius 10 cm is cut into two pieces and the pieces are bent round to form two circles. Find the lengths of the two pieces when the sum of the areas of the circles is as small as possible.

9. A tank has a horizontal square base and vertical rectangular sides but is open at the top. It is built to contain 20 ft^3 of water when full. Find the ratio of the height to the edge of the base when the total area of the sides and base is a minimum.

10. A straight line drawn through the point (8, 1) cuts the axes at *P* and *Q*. Assuming that *PQ* has a negative gradient, find the minimum length of *PQ*.

Small changes

If $y = f(x)$ we have

$$D_x y = \lim_{\Delta x \to 0} \frac{\Delta y}{\Delta x}.$$

When Δx is small,

$$\frac{\Delta y}{\Delta x} \approx D_x y,$$

and so

$$\Delta y \approx (D_x y) \,.\, \Delta x.$$

EXAMPLE. *If the radius of a sphere increases from* 10 *cm to* 10·1 *cm, find an approximation for the increase in the area of the surface of the sphere.*

Let the radius of the sphere be r cm and the area A cm^2; then $A = 4\pi r^2$. Hence

$$D_r A = 8\pi r, \quad \text{and} \quad \Delta A \approx 8\pi r \,.\, (\Delta r).$$

But $r = 10$ and $\Delta r = 0{\cdot}1$, so

$$\Delta A \approx 8\pi \,.\, 10 \,.\, 0{\cdot}1 = 8\pi.$$

Hence the increase in the area of the sphere is approximately 8π cm^2.

EXAMPLE. *The time of a complete oscillation of a simple pendulum is given by $t = 2\pi\sqrt{(l/g)}$. If the length of a pendulum with a period of oscillation of 2 sec is increased by 3 per cent, what is the percentage increase in the period?*

If t sec is the period of oscillation of a pendulum of length l metres,

$$t = 2\pi\sqrt{\left(\frac{l}{g}\right)},$$

where g is constant. Then

$$D_l t = \frac{\pi}{\sqrt{(lg)}}, \qquad \text{and} \qquad \Delta t \approx \frac{\pi}{\sqrt{(lg)}} \,.\, \Delta l.$$

Hence

$$\frac{\Delta t}{t} \approx \frac{1}{2} \,.\, \frac{\Delta l}{l}.$$

But

$$\frac{\Delta l}{l} = \frac{3}{100},$$

so that

$$\frac{\Delta t}{t} \approx \frac{1{\cdot}5}{100}.$$

Therefore the period of the pendulum is increased by approximately 1·5 per cent.

EXERCISE 12.3

1. If the radius of a circle is increased from 5 cm to 5·1 cm, find the approximate increase in the area of the circle.

2. If the radius of a sphere is increased from 10 cm to 10·2 cm, find an approximation for the increase in the volume of the sphere.

3. A sphere of metal is heated so that its radius is increased by 1·5 per cent. Find the approximate percentage increase in the area of the surface of the sphere.

4. The pressure p gm/cm^2 and the volume v cm^3 of a quantity of gas contained in a spherical bubble are connected by the relation $pv^{1\cdot 4} = c$, where c is a constant. Find the approximate percentage decrease in the pressure when the bubble expands so that its radius is increased by 2 per cent.

5. A simple pendulum is moved from A to B, and the value of g at B is 0·5 per cent lower than it was at A. Find the approximate percentage change in the period of oscillation of the pendulum caused by the move. [$t = 2\pi\sqrt{(l/g)}$.]

6. The intensity of the magnetic field due to a current of A amp in a wire in the form of a circle of radius a cm, at a point on the axis of the circuit and distant x cm from the centre of the circle, is $\frac{1}{5}\pi Aa^2(x^2 + a^2)^{-3/2}$ dynes. If $a = 3$ and $A = 50$, find approximately the increase in the strength of the field when x is changed from 4 to 3·9 cm.

7. An object P on the axis of a spherical mirror of radius r cm has an image Q on the axis of the mirror. If the distances of P and Q from the mirror are respectively u cm and v cm,

$$\frac{1}{v} + \frac{1}{u} = \frac{2}{r}.$$

The linear magnification caused by the reflexion is defined to be $\Delta v/\Delta u$. Find an approximation for the linear magnification in terms of u and v.

8. A tinned goods manufacturer uses a cylindrical can with a fixed capacity of 40π cm^3. Find an approximation for the increase in the total surface area of the can when the radius of the base is increased from 4 cm to 4·1 cm.

Differential equations

Nearly all mathematical operations have a useful inverse operation; for example, squaring and finding the square root, or looking up a logarithm and finding the antilogarithm. Differentiation is no exception, and the inverse operation is called *integration*; the reason for this name will appear later.

The problem now is: given the derived function, find the original function, sometimes called the *primitive*.

Suppose $D_xy = 3x^2$, then we could have $y = x^3$, but we could also have $y = x^3 + 1$, $y = x^3 - 2$, and so on. An equation like $D_xy = 3x^2$ is called a *differential equation*. We see that its solution is $y = x^3 + c$, where c is a constant; the latter is called the *constant of integration*.

EXAMPLE. *Solve the differential equation* $D_x y = x^2 + 2x - 3$, *given that* $y = 0$ *when* $x = -1$.

[If we differentiate x^3, we get $3x^2$; if we differentiate $\frac{1}{3}x^3$, we get x^2; the integral of x^2 is therefore $\frac{1}{3}x^3$.]

Since

$$D_x y = x^2 + 2x - 3,$$

$$y = \tfrac{1}{3}x^3 + 2(\tfrac{1}{2}x^2) - 3x + c.$$

But

$$y = 0 \quad \text{when } x = -1;$$

$$0 = -\tfrac{1}{3} + 1 + 3 + c, \quad \text{or} \quad c = -\tfrac{11}{3}.$$

Then the solution is

$$y = \tfrac{1}{3}x^3 + x^2 - 3x - \tfrac{11}{3}.$$

Ex. 2. Solve the differential equations

(i) $D_x y = 4x^3 - 6x + 3$, given that $y = 1$ when $x = 1$;

(ii) $D_x y = \dfrac{1}{x^3} - \dfrac{2}{x^2} + 3x$, given that $y = 1\frac{1}{2}$ when $x = \frac{1}{2}$.

EXAMPLE. *The gradient of a curve at a point* (x, y) *is given by* $(8x - 3)$ *and the curve passes through the point* $(1, 8)$. *Find the equation of the curve.*

We have

$$D_x y = 8x - 3,$$

giving

$$y = 4x^2 - 3x + c.$$

But

$$y = 8 \text{ when } x = 1,$$

so that

$$8 = 4 - 3 + c.$$

Hence

$$y = 4x^2 - 3x + 7$$

is the equation of the curve.

EXAMPLE. *The speed* v *ft/sec of a stone thrown vertically upwards, after* t *seconds when it has travelled* x *feet, is given by* *the formula $v = 90 - 32t$. At what height does the stone come instantaneously to rest?*

We are given that

$$D_t x = 90 - 32t.$$

Then

$$x = 90t - 16t^2 + k.$$

But

$$x = 0 \text{ when } t = 0 \Rightarrow k = 0,$$

so

$$x = 90t - 16t^2.$$

Now the speed is zero when $90 - 32t = 0$, or when $t = 45/16$; then

$$x = 90(\tfrac{45}{16}) - 16(\tfrac{45}{16})^2 = \tfrac{45^2}{16}.$$

Hence the stone comes instantaneously to rest at a height of just over $126\frac{1}{2}$ ft.

EXERCISE 12.4

1. (i) Differentiate x^2 and integrate x; (ii) Differentiate x^3 and integrate x^2; (iii) Differentiate $1/x$ and integrate $1/x^2$; (iv) Differentiate $\sqrt{x}$ and integrate $1/\sqrt{x}$.

2. Write down the integrals (or primitive functions) of:

(i) 4; (ii) $2x^2$; (iii) $3x + 2$; (iv) x^4; (v) $-\dfrac{2}{x^2}$; (vi) $-\dfrac{1}{x^2} + \dfrac{2}{x^3}$;

(vii) $x^2 + \dfrac{1}{x^2}$; (viii) $3x^{1/2}$; (ix) $x^2 - 2x + 3$; (x) $\sqrt{x} + \dfrac{1}{\sqrt{x}}$;

(xi) $\dfrac{2x - 1}{\sqrt{x}}$; (xii) $\dfrac{3x - 2}{x^3}$.

3. Solve the following differential equations, using the given conditions to determine the constant of integration:

(i) $D_x y = \frac{1}{3}x^2 + \frac{1}{2}x - 3/x^2$, $y = 34$ when $x = 6$;

(ii) $D_x y = \dfrac{5x^2 - 3x + 2}{\sqrt{x}}$, $y = 20$ when $x = 4$.

4. Find the equation of the curve which passes through the point (2, 5) and is such that, at the point (x, y),

$$D_x y = 3x^2 - 8x + 4.$$

5. A curve passes through the origin and its gradient function is $(2x - 3x^2)$. Find the ordinate when $x = 2$.

6. A particle moves along a straight line OX, its distance from O at time t sec being s ft and its velocity v ft/sec at that time being given by $v = 9t^2 + 4t + 7$. Find the distance travelled between $t = 3$ and $t = 10$.

7. The speed of a car over a certain stretch of road is v ft/sec at time t sec, where $v = 25 + 3{\cdot}5t$. Find the distance it travels in the 6 sec after reaching the speed of 25 ft/sec.

8. A curve passes through the point $(1, -1)$ and the gradient of the tangent at any point (x, y) on the curve is $\frac{3}{10}(x^2 + 2x - 8)$. Find the equation of the curve and the coordinates of its turning points.

9. A particle starts from rest at O and moves along a straight line. After t seconds its velocity is v cm/sec, where $v = 3t^2 - t^3$. Show that the particle is momentarily at rest after 3 sec and find its distance from O at this time.

Find also the maximum velocity of the particle during the first 3 sec of the motion.

Velocity and acceleration

Suppose a particle is moving along a straight line OX. It passes a fixed point O on the line, and t seconds afterwards it has reached a position P, where $OP = x$ metres. The velocity of the particle at P is $D_t x$ m/sec. Velocity is the rate of change of the distance OP and we denote it by v metres/sec.

The name *acceleration* is given to the rate of change of the velocity at P; it is measured by $D_t v$ and we shall denote it by a m/sec per sec, or a m/sec^2.

EXAMPLE. *If the acceleration of the particle along OX is constant, find x in terms of t.*

Let

$$a = f, \text{ a constant.}$$

Then

$$D_t v = f,$$

$$v = ft + g, \text{ where } g \text{ is a constant.}$$

Suppose

$$v = u \text{ when } t = 0 \quad \Rightarrow \quad u = g,$$

$$v = u + ft,$$

$$D_t x = u + ft,$$

$$x = ut + \tfrac{1}{2}ft^2 + h.$$

But

$$x = 0 \text{ when } t = 0 \ \Rightarrow \ h = 0,$$

so

$$x = ut + \tfrac{1}{2}ft^2.$$

$$\boldsymbol{a = v\, D_s v = \tfrac{1}{2}D_s(v^2)}$$

The chain rule for differentiation gives

$$a = D_t v = \mathbf{D_s v} \,.\, D_t s$$

where t, s, v, a are respectively the time, distance, velocity and acceleration of a particle moving along a straight line.

But $v = D_t s$, and $D_s(v^2) = 2vD_s v$, so that

$$a = D_s v \,.\, v = \tfrac{1}{2}D_s(v^2).$$

This formula for acceleration is useful because it is often convenient to express velocity in terms of distance; for example, the ordinary speedometer gives readings of distance and speed, but not usually of time.

EXAMPLE. *A body moves in a straight line with constant acceleration f m/sec^2, the initial velocity being u m/sec. When the body has travelled a distance s metres its speed is v m/sec. Express v in terms of s.*

Let $a = f$, a constant.

$$D_s(v^2) = 2f,$$

$$v^2 = 2fs + k, \text{ where } k \text{ is a constant.}$$

But

$$v = u \text{ when } s = 0 \Rightarrow u^2 = k,$$

and so

$$v^2 = u^2 + 2fs.$$

EXERCISE 12.5

1. A body moves in a straight line in such a way that its acceleration $a = 3t$, the units of distance and time (measured from an origin O) being metres and seconds. Find its distance from O after 5 sec, given that its speed when it passes O is 10 m/sec.

2. A particle moves along a straight line so that, after t sec, $a = t + 2$ where a cm/sec^2 is the acceleration. If the particle started from rest, find its velocity at the end of 3 sec and the distance covered during the next second.

3. The retardation (negative acceleration) of a particle moving in a straight line is $4x$ m/sec^2 when the particle is x metres from a fixed point O on the line. The velocity when $x = 2$ is 3 m/sec. Find the velocity when $x = 0$ and the greatest positive value of x.

4. A small body attached to the end of an elastic string is allowed to fall from rest and has an acceleration of $(32 - 2x)$ ft/sec^2 after it has fallen x feet. Find its greatest velocity and the distance it falls before first coming to rest.

5. A particle is moving in a straight line. It moves from rest at a fixed point O and t seconds later its distance from O is $t^2(2t - 5)$ feet. Find (i) its acceleration at the start, (ii) its distance from O when it next comes to rest, (iii) its average velocity during the first 3 sec of its motion.

6. A particle travels in a straight line, being at a point A at zero time. Its velocity in metres per second at time t seconds is given by $v = 6t^2 - 48t + 42$. Find expressions at time t sec for (i) its acceleration, and (ii) its distance from A.

Find the times at which its velocity is zero. Show also that it passes through A twice after zero time, and find (to one place of decimals) the length of time before its first return.

7. A body starts from rest and moves in a straight line for the first 5 sec with an acceleration of $(30 - 6t)$ ft/sec^2, where t is the time in seconds, and subsequently with a constant velocity. Find (i) the magnitude of the constant velocity, and (ii) the distance travelled in the first 5 sec.

Write down an expression for the total distance travelled during the first $(k + 5)$ sec of the motion, and deduce the value of k at which the average velocity will have reached 70 ft/sec.

8. A particle starts from rest at a point 6 ft from O, and moves in a straight line away from O with a velocity v ft/sec at time t sec given by $v = t - \frac{1}{18}t^2$. Find

(i) its acceleration and distance from O in terms of t;

(ii) the time at which it begins to return, and the time at which it again reaches its starting point.

9. What does the relation $D_x y = D_u y . D_x u$ become when $y = x$? Show that $D_x t = v^{-1}$.

When a meteorite comes into the earth's gravitational field, its speed towards the earth is v k.p.h. when it is x kilometres from the earth's centre. Assume that $v^2 \approx c^2/x$, where c is a constant, and so find (i) its acceleration when $x = a$, (ii) the time taken while x decreases from a to b.

10. A particle which is repelled from a point O with an acceleration $3k^2\sqrt{x}$ cm/sec when it is x cm from O, is placed in unstable equilibrium at O. If the particle is slightly disturbed, find its velocity when it is x cm from O, and the time taken to go that distance.

CHAPTER 13

FURTHER DIFFERENTIATION AND APPLICATIONS

Differentiation of a product

Suppose

$$y = (3x + 7)(x^2 + 9),$$

$$y + \Delta y = \{3(x + \Delta x) + 7\}\{(x + \Delta x)^2 + 9\},$$

$$\Delta y = \{(3x + 7) + 3\Delta x\}\{(x^2 + 9) + 2x(\Delta x) + (\Delta x)^2\} - (3x + 7)(x^2 + 9)$$

$$= (3x + 7)\{2x(\Delta x) + (\Delta x)^2\} + (x^2 + 9)\{3\Delta x\} + \{3\Delta x\}\{2x(\Delta x) + (\Delta x)^2\},$$

$$\frac{\Delta y}{\Delta x} = (3x + 7)(2x + \Delta x) + (x^2 + 9)(3) + (3)\{2x(\Delta x) + (\Delta x)^2\}.$$

$$D_x y = \lim_{\Delta x \to 0} (\Delta y/\Delta x)$$

$$= (3x + 7) . 2x + (x^2 + 9) . 3 + 3 . 0$$

$$= 9x^2 + 14x + 27.$$

This method of differentiating the product of two functions can be applied generally.

Suppose $y = uv$, where u and v are functions of x:

$$y + \Delta y = (u + \Delta u)(v + \Delta v),$$

$$\Delta y = (u + \Delta u)(v + \Delta v) - uv$$

$$= u . \Delta v + v . \Delta u + \Delta u . \Delta v,$$

$$\frac{\Delta y}{\Delta x} = u \, . \frac{\Delta v}{\Delta x} + v \, . \frac{\Delta u}{\Delta x} + \frac{\Delta u}{\Delta x} \, . \, \Delta v.$$

$$\begin{aligned} \mathrm{D}_x y &= \lim_{\Delta x \to 0} (\Delta y/\Delta x) \\ &= u \, . \, \mathrm{D}_x v + v \, . \, \mathrm{D}_x u + (\mathrm{D}_x u) \, . \, 0. \end{aligned}$$

Hence

$$\mathbf{D}_x(\boldsymbol{uv}) = \boldsymbol{u} \, . \, \mathbf{D}_x \boldsymbol{v} + \boldsymbol{v} \, . \, \mathbf{D}_x \boldsymbol{u}.$$

Differentiation of a quotient

Suppose $y = u/v$, where u and v are functions of x:

$$y + \Delta y = \frac{u + \Delta u}{v + \Delta v},$$

$$\begin{aligned} \Delta y &= \frac{u + \Delta u}{v + \Delta v} - \frac{u}{v} \\ &= \frac{v \, . \, \Delta u - u \, . \, \Delta v}{v(v + \Delta v)}, \end{aligned}$$

$$\frac{\Delta y}{\Delta x} = \frac{v \, . \, \dfrac{\Delta u}{\Delta x} - u \, . \, \dfrac{\Delta v}{\Delta x}}{v(v + \Delta v)}.$$

$$\begin{aligned} \mathrm{D}_x y &= \lim_{\Delta x \to 0} (\Delta y/\Delta x) \\ &= \frac{v \, . \, \mathrm{D}_x u - u \, . \, \mathrm{D}_x v}{v^2}. \end{aligned}$$

Hence

$$\mathbf{D}_x\left(\frac{\boldsymbol{u}}{\boldsymbol{v}}\right) = \frac{\boldsymbol{v} \, . \, \mathbf{D}_x \boldsymbol{u} - \boldsymbol{u} \, . \, \mathbf{D}_x \boldsymbol{v}}{\boldsymbol{v}^2}.$$

In the differentiation of uv and u/v the following theorems on limits have been used:

$$\lim_{x \to a} \{f(x) \pm g(x)\} = \lim_{x \to a} f(x) \pm \lim_{x \to a} g(x);$$

$$\lim_{x \to a} f(x) \, . \, g(x) = \lim_{x \to a} f(x) \, . \lim_{x \to a} g(x);$$

$$\lim_{x \to a} \frac{f(x)}{g(x)} = \frac{\lim\limits_{x \to a} f(x)}{\lim\limits_{x \to a} g(x)}, \quad \text{if } \lim_{x \to a} g(x) \neq 0.$$

Proofs of these theorems are the concern of more advanced analysis than will be considered in this book, but will be found in books on analysis written for university students.

EXAMPLE. *Differentiate* (*i*) $x^2\sqrt{(1 + x)}$, (*ii*) $\dfrac{x^2}{\sqrt{(1 + x)}}$.

$$Dx^2 = 2x, \quad D\sqrt{(1 + x)} = \frac{1}{2\sqrt{(1 + x)}}.$$

(i) $$D\{x^2\sqrt{(1 + x)}\} = x^2 . \frac{1}{2\sqrt{(1 + x)}} + \sqrt{(1 + x)} . 2x$$

$$= \frac{x^2 + 4x(1 + x)}{2\sqrt{(1 + x)}}$$

$$= \frac{x(5x + 4)}{2\sqrt{(1 + x)}}.$$

(ii) $$D\left(\frac{x^2}{\sqrt{(1 + x)}}\right) = \frac{\sqrt{(1 + x)} . 2x - x^2 . \dfrac{1}{2\sqrt{(1 + x)}}}{(1 + x)}$$

$$= \frac{4x(1 + x) - x^2}{2(1 + x)^{3/2}}$$

$$= \frac{x(3x + 4)}{2(1 + x)^{3/2}}.$$

It is sometimes more convenient to write u/v as uv^{-1}, and to differentiate uv^{-1} as a product:

$$D(uv^{-1}) = u . (-v^{-2} . Dv) + v^{-1} . Du$$

$$= \frac{1}{v^2}(v . Du - u . Dv).$$

Ex. 1. Differentiate (i) $(1 - x)^2\sqrt{(x + 2)}$, (ii) $\dfrac{\sqrt{(x + 2)}}{(1 - x)^2}$.

Implicit differentiation

Suppose

$$z = xy,$$

then

$$D_x z = x \,.\, D_x y + y \,.\, 1,$$

or

$$D_x(xy) = xD_x y + y.$$

We saw on p. 167 that $D_x(y^2) = 2y \,.\, (D_x y)$ and so equations like $x^2 + 2xy - 3y^2 + 5x - y - 15 = 0$ can now be differentiated term by term.

EXAMPLE. *Find the equation of the tangent to the curve* $x^2 + 2xy - 3y^2 + 5x - y - 15 = 0$ *at the point* (2, 1).

Differentiating term by term with respect to x:

$$2x + 2(x \,.\, Dy + y) - 6y \,.\, Dy + 5 - Dy = 0.$$

If m is the gradient of the curve at (2, 1),

$$4 + 4m + 2 - 6m + 5 - m = 0,$$

giving

$$m = \tfrac{11}{3}.$$

The tangent to the curve at (2, 1) is therefore

$$y - 1 = \tfrac{11}{3}(x - 2)$$

or

$$11x - 3y = 19.$$

EXERCISE 13.1

1. Differentiate $(2x + 3)(x - 2)$ as a product and check by differentiating $2x^2 - x - 6$.

2. Differentiate $\dfrac{x}{x^2 + 1}$ as a quotient and check by differentiating $\left(x + \dfrac{1}{x}\right)^{-1}$.

3. Differentiate the products:

(i) $x^2(x - 1)^2$; (ii) $x\sqrt{(x + 1)}$; (iii) $(x + 2)^2(x - 1)^2$;
(iv) $(2x - 1)^2(3x + 1)^3$; (v) $(2x + 1)\sqrt{(x^2 + 2)}$; (vi) $x^n(2 - x)^n$.

4. Differentiate the quotients:

(i) $\dfrac{3x + 2}{2 - 3x}$; (ii) $\dfrac{x^2 + 2}{2x - 1}$; (iii) $\dfrac{\sqrt{(1 + x)}}{x}$;

(iv) $\dfrac{a - bx}{c + dx}$; (v) $\dfrac{\sqrt{(x^2 + 2)}}{x}$; (vi) $\dfrac{\sqrt{x}}{\sqrt{(x + 1)}}$.

5. Show that the function $x/(x^2 - x - 2)$ is a decreasing function of x for all values of x.

6. Find the stationary values of $(2x + 1)/(x^2 - 8x - 2)$, distinguishing between maxima and minima.

7. Find the equations of the tangents to the curve $2x^2 + 3xy - 2y^2 = 12$ at the points where $x = 2$.

8. Differentiate $xy + ax + c = 0$ and hence show that $x^2\text{D}y$ is constant.

9. Find the equations of the tangent and normal to the curve $x^3 + y^3 = 9xy$ at the point (4, 2).

10. If $y = \left(\dfrac{x}{x + a}\right)^n$, prove that $\text{D}y = \dfrac{nay}{x(x + a)}$.

11. If $y = \left(\dfrac{x^2 + 1}{x^2 + 2}\right)^n$, prove that $\text{D}y = \dfrac{2nxy}{(x^2 + 1)(x^2 + 2)}$.

12. Solve the differential equation $x\text{D}y = x^3 + y$ by means of the substitution $y = tx$, where t is a variable.

13. Solve the differential equation $(x + y)(x\text{D}y - y) = x^4$ by putting $y = vx$.

14. Find the stationary values of the following functions, stating the values of x for which they occur, and distinguishing between maxima and minima. Sketch the graphs:

(i) $(x - 2)^2(x - 1)$; (ii) $\dfrac{x(x + 1)}{x - 1}$;

(iii) $\dfrac{x\sqrt{(1 - x)}}{2x + 1}$; (iv) $\dfrac{\sqrt{(x + 1)}}{x^2}$.

15. Show that the function $\dfrac{(x + 2)\sqrt{(x + 1)}}{(x + 3)^{3/2}}$ has no stationary values.

The second derivative

Suppose y is a function of x, say $y = f(x)$. The derived function, or derivative, Dy, is itself a function of x. If we differentiate again, we obtain the second derived function of y, or the second derivative; this can be denoted by D^2y, or $f''(x)$.

Dy measures the rate of change of y with respect to x, or the gradient of the graph of y against x at the point (x, y).

D^2y measures the rate of change of Dy with respect to x, or the gradient of the graph of Dy against x at the point (x, Dy).

Figure 13.1 shows how the graphs of y, Dy and D^2y are related. A study of corresponding points on the three curves, such as A, A_1 and A_2, brings to light the following properties of y, Dy and D^2y.

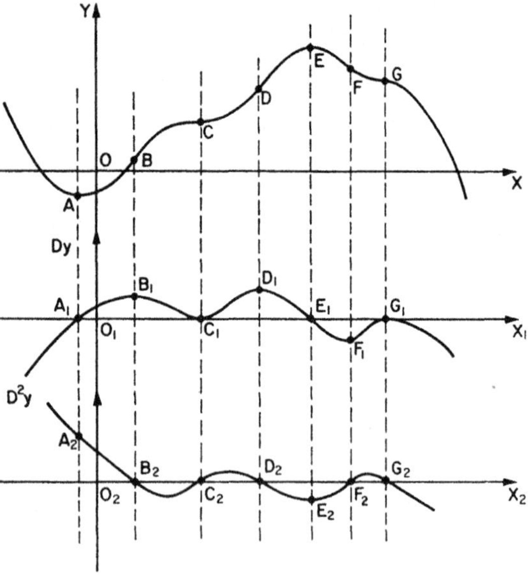

Fig. 13.1

(i) At a minimum point of $y = f(x)$, the value of the derived function $f'(x)$ is zero and the second derived function $f''(x)$ is positive.

(ii) At a maximum point of $y = f(x)$, the value of the derived function $f'(x)$ is zero and the second derived function $f''(x)$ is negative.

(iii) At a point of inflection of $y = f(x)$, the derived function $f'(x)$ is either maximum or minimum; the second derivative $f''(x)$ is zero *and changes sign.*

(iv) When the graph of $y = f(x)$ is *concave upwards*, as in the arcs AB, CD, FG in Fig. 13.1, the second derivative $f''(x)$ *is positive*; when the graph of $y = f(x)$ is *concave downwards*, as in the arcs BC, DF in Fig. 13.1, the second derivative $f''(x)$ is negative. A point of inflection is a point at which the concavity is reversed.

The sign of the second derivative can therefore be used to distinguish between maxima and minima of the primitive function, and the zeros of $f''(x)$, provided they mark a change of sign of $f''(x)$, serve to locate the points of inflection on the graph of $y = f(x)$.

The sign of $f''(x)$ is more commonly used in investigating the stationary values of trigonometrical and other functions; with rational functions it is often easier to examine the changes of sign of $f'(x)$. The following Example gives a comparison of the two methods and shows how the evaluation of D^2y can be simplified when $Dy = 0$.

EXAMPLE. *Show that the maximum value of* $\sqrt{x}/(1+x)$ *is* $\frac{1}{2}$.

Let

$$y = \frac{\sqrt{x}}{1+x}.$$

$$Dy = \left\{\frac{1}{2\sqrt{x}}\,.\,(1+x) - 1\,.\,\sqrt{x}\right\}\Big/(1+x)^2$$

$$= \left\{\frac{(1+x) - 2x}{2\sqrt{x}}\right\}\Big/ (1+x)^2,$$

$$Dy = -\frac{(x-1)}{2\sqrt{x}(1+x)^2}.$$

$Dy = 0$ only when $x = 1$. To evaluate D^2y when $x = 1$ we write

$$Dy = -(x - 1) \,.\, u \quad \text{where } u = \frac{1}{2\sqrt{x}(1 + x)^2}.$$

Then

$$D^2y = -u - (x - 1)Du;$$

when $x = 1$, $\qquad D^2y = -\frac{1}{8} - 0 < 0.$

Also, when $x = 1$, $y = \frac{1}{2}$.

Hence y has a maximum value of $\frac{1}{2}$ when $x = 1$.

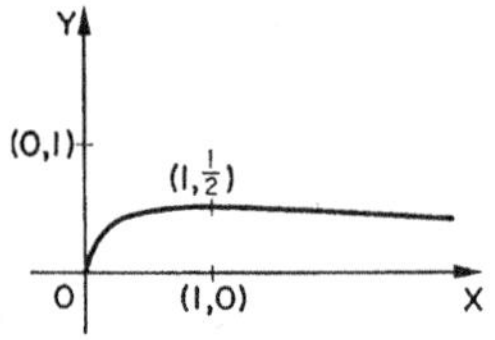

Fig. 13.2

Figure 13.2 shows the graph of

$$y = \frac{\sqrt{x}}{1 + x}.$$

The domain is $x \geqslant 0$ since $\sqrt{x}$ does not exist when $x < 0$. Also, when x is large, $y \approx 1/\sqrt{x}$ and $y \to 0$ as $x \to \infty$. The axis of x is an *asymptote* of the curve.

Note. In the above Example we have

$$Dy = -\frac{(x - 1)}{2\sqrt{x}(1 + x)^2}.$$

Clearly the value of Dy changes from $+$ to $-$ as x passes through the value $x = 1$. Hence $(1, \frac{1}{2})$ is a maximum point. The value of D^2y at $x = 1$ is not essential to distinguish between maxima and minima; it was used above to illustrate a method which will have more general value later on.

EXAMPLE. *Find the points of inflection on the graph of* $y = x^4 - 2x^3 - 12x^2 + 36x - 12.$

$$Dy = 4x^3 - 6x^2 - 24x + 36,$$

$$D^2y = 12x^2 - 12x - 24 = 12(x + 1)(x - 2).$$

$D^2y = 0$ *and changes sign* when $x = -1$ or 2. The points of inflection on the graph are therefore $(-1, -57)$ and $(2, 12)$.

EXAMPLE. *Examine the stationary values of the function* $y = (x - 1)(x - 3)^3$, *and sketch its graph.*

$$Dy = 1 \,.\, (x - 3)^3 + 3(x - 3)^2 \,.\, (x - 1)$$

$$= 2(2x - 3)(x - 3)^2.$$

$$D^2y = 4 \,.\, (x - 3)^2 + 4(x - 3) \,.\, (2x - 3)$$

$$= 12(x - 2)(x - 3).$$

When $x = \frac{3}{2}$, $y = -\frac{27}{16}$, $Dy = 0$, $D^2y = 9$;

$x = 2$, $y = -1$, $Dy = 2$, $D^2y = 0$ and changes from positive to negative;

$x = 3$, $y = 0$, $Dy = 0$, $D^2y = 0$ and changes from negative to positive.

$(\frac{3}{2}, -\frac{27}{16})$ is a minimum; $(2, -1)$ is an inflection; $(3, 0)$ is a stationary inflection. Changes of sign help in sketching the curve:

$y < 0$ when $1 < x < 3$; $y > 0$ when $x < 1$ and when $x > 3$;

$Dy < 0$ when $x < \frac{3}{2}$; $Dy \geqslant 0$ when $x \geqslant \frac{3}{2}$;

$D^2y < 0$ when $2 < x < 3$; $D^2y > 0$ when $x < 2$ and when $x > 3$.

The function is a quartic polynomial; it is continuous for all values of x and is single valued. A sketch of the graph is shown in Fig. 13.3.

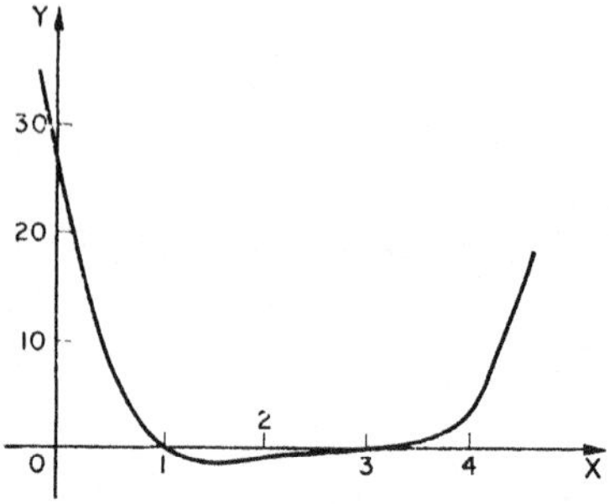

Fig. 13.3

EXAMPLE. *Find* $D_x y$ *and* $D_x^2 y$ *in terms of t when*

$$x = \frac{t}{1+t}, \quad y = \frac{t^2}{1+t}.$$

$$D_t x = \frac{1.(1+t) - 1.t}{(1+t)^2} = \frac{1}{(1+t)^2},$$

$$D_t y = \frac{2t(1+t) - 1.t^2}{(1+t)^2} = \frac{t(2+t)}{(1+t)^2}.$$

But

$$D_t y = D_x y \,.\, D_t x \;\Rightarrow\; D_x y = \frac{D_t y}{D_t x}.$$

Hence

$$D_x y = t(2+t).$$

Let

$$p = D_x y;$$

then

$$D_x p = D_x^2 y.$$

But

$$D_x p = D_t p / D_t x,$$
$$p = 2t + t^2,$$
$$D_t p = 2(1+t).$$

Therefore

$$D_x^2 y = 2(1+t)/(1+t)^{-2},$$

and so

$$D_x^2 y = 2(1+t)^3.$$

Note. When x and y are the coordinates of a point on the graph of $y = f(x)$, and are expressed in terms of a third variable t, the variable t is called the *parameter* of the point (x, y). If t is eliminated the cartesian or x, y equation of the curve is found. In the Example above, $y = tx$ or $t = y/x$, and so

$$x = \frac{y}{x + y}, \quad \text{or} \quad x^2 + xy - y = 0.$$

EXERCISE 13.2

1. Find the maximum and minimum values of the following polynomials, using the second derivative to distinguish between them; also find the points of inflection of their graphs and sketch the graphs:
(i) $2x^3 + 3x^2 - 12x + 7$; (ii) $3 + 72x + 60x^2 - 16x^3$;
(iii) $3x^4 + 8x^3 - 30x^2 - 72x - 27$; (iv) $2x^4 + 8x^3 + 5x^2 - 12x - 12$.

2. Find the equations, $y = f(x)$, of the curves with the following properties:
(i) $f(x)$ is a cubic polynomial, there is a point of inflection at $(\frac{1}{2}, 0)$, the gradient when $x = 0$ is -1, and the curve passes through $(0, 1)$;
(ii) $f(x)$ is a quartic polynomial, there are points of inflection when $x = -\frac{1}{2}$ and when $x = 2$, there is a minimum at $(1, -2)$ and the gradient when $x = 2$ is 11.
(iii) $f(x)$ is a rational function of the form

$$\frac{ax + b}{cx^2 + d},$$

with a minimum at $(-1, -\frac{1}{2})$ and a maximum at $(1, \frac{1}{2})$; also find the points of inflection.

3. Given that $f'(x) = (x - a)F(x)$ and $F(a) < 0$, prove that $f(x)$ has a maximum value when $x = a$.
Find the coordinates of the stationary points of the curve

$$y = \frac{3 + 2x}{4 + x^2},$$

distinguishing between maximum and minimum. Sketch the graph of the function.

4. Express $D_x y$ and $D_x^2 y$ in terms of t when

$$x = \frac{2 + t}{1 + 2t}, \quad y = \frac{3 + 2t}{t}.$$

5. If $x = \dfrac{t}{t + 1}$, $y = \dfrac{2t + 1}{(t + 1)^2}$ find $D_x y$ in terms of t and prove that $D_x^2 y$ is constant.

6. If $x = 1 - t^2$, $y = t - t^3$, find $D_x y$ and $D_x^2 y$ in terms of t. Hence show that y has a maximum value of $\frac{2}{9}\sqrt{3}$ when $x = \frac{2}{3}$.

7. Prove that, when $y = at^2 + 2bt + c$, $t = ax^2 + 2bx + c$, and a, b, c are constants, $D_x^3 y = 24a^2(ax + b)$. [$D_x^3 y$ is the derivative of $D_x^2 y$ with respect to x.]

8. A right circular cone of variable height h is inscribed in a sphere of radius R. Show that the volume of the cone is a maximum when $h = 4R/3$.

9. Prove that the maximum volume of a right circular cylinder inscribed in a sphere of radius R is $4\pi R^3\sqrt{3}/9$.

10. A sphere of radius a rests on a table. A hollow cone rests on the sphere and its base rests on the table. If the base radius of the cone is x, show that its volume is $\frac{2}{3}\pi a x^4/(x^2 - a^2)$. Hence show that the minimum volume of the cone is twice the volume of the sphere.

Acceleration

Suppose a particle is moving in a straight line; t seconds after passing a fixed point O in the line, the particle is at a point P where $OP = s$ feet. The velocity when the particle is at P is v ft/sec and the acceleration is a ft/sec². We have seen on p. 182 that $a = D_t v = vD_s v = \frac{1}{2}D_s(v^2)$.

But $v = D_t s$, so $a = D_t v = D_t^2 s$. We therefore have the following expressions for a:

$$D_t v, \quad D_t^2 s, \quad vD_s v, \quad \tfrac{1}{2}D_s(v^2).$$

EXAMPLE. *If $s = 2t^3 - 15t^2 + 36t + 20$, find the times when the velocity and acceleration are zero and the initial acceleration.*

$$v = D_t s = 6t^2 - 30t + 36 = 6(t - 2)(t - 3),$$

$$a = D_t v = 12t - 30 = 6(2t - 5).$$

Hence the velocity is zero when $t = 2$ and when $t = 3$, the acceleration is zero when $t = 2\frac{1}{2}$ and the initial acceleration is -30 ft/sec².

$D_x y = 1/D_y x$

If $y = f(x)$ and $x = g(u)$, the chain rule gives

$$D_u y = D_x y \,.\, D_u x.$$

When $y = u$, $\quad y = f(x)$ and $x = g(y)$.

The functions f and g are therefore *inverse functions*. For example,

$$y = x^2 \Rightarrow x = \sqrt{y}$$

$$y = \frac{x}{1+x} \Rightarrow x = \frac{y}{1-y}.$$

The chain rule then becomes, since $D_u y = 1$,

$$1 = D_x y \,.\, D_y x$$

and so

$$D_x y = \frac{1}{D_y x}.$$

EXAMPLE. *A particle moves in a line OA so that its velocity at time t seconds is* $\frac{1}{3}\sqrt{(s+2)}$ *ft/sec, where s feet is its distance from O. Find the time taken for the particle to increase its distance from O from* 2 *ft to* 14 *ft.*

$$v = \tfrac{1}{3}\sqrt{(s+2)}, \text{ or } D_t s = \tfrac{1}{3}\sqrt{(s+2)}.$$

Hence

$$D_s t = \frac{3}{\sqrt{(s+2)}},$$

and so

$$t = 6\sqrt{(s+2)} + c.$$

Suppose $t = t_1$ when $s = 2$; then $c = t_1 - 12$. If $t = t_2$ when $s = 14$, $t_2 = 24 + (t_1 - 12)$.

Therefore

$$t_2 - t_1 = 12.$$

Hence the required time is 12 sec.

EXAMPLE. *A projectile moves horizontally with a retardation of* $10^{-7}v^3$ *ft/sec when its velocity is v ft/sec. Find the distance it travels while its velocity is reduced from* 4000 *to* 3600 *ft/sec, and the time taken.*

We are given that $v \,.\, D_s v = -\dfrac{v^3}{10^7}$. Then

$$D_s v = -\frac{v^2}{10^7}, \text{ so } D_v s = -\frac{10^7}{v^2}.$$

Hence

$$s = \frac{10^7}{v} + c.$$

Suppose $s = 0$ when $v = 4000$, so that $c = -\frac{10^7}{4000}$.

When $v = 3600$, $s = 10^7(\frac{1}{3600} - \frac{1}{4000})$, and so the required distance is

$$10^5(\tfrac{1}{36} - \tfrac{1}{40}) \text{ ft} = \frac{10^5 \times 4}{36 \times 40} \text{ ft},$$

which is approximately 278 ft.

Now $s = 10^7\left(\frac{1}{v} - \frac{1}{4000}\right)$, giving

$$\frac{10^7}{v} = s + 2500.$$

But

$$v = D_t s, \quad \text{and so} \quad \frac{1}{v} = D_s t.$$

Therefore

$$10^7 D_s t = s + 2500,$$

$$10^7 \,.\, t = \tfrac{1}{2}s^2 + 2500s + k.$$

Taking $t = 0$ when $s = 0$, then $k = 0$. The required time is given by putting

$$s = \frac{10^5 \times 4}{36 \times 40};$$

then

$$10^7 \,.\, t = \frac{10^5 \times 2}{36 \times 40}\left\{\frac{10^5 \times 4}{36 \times 40} + 5000\right\},$$

so that the time is approximately 0·073 sec.

The Leibniz notation

If y is a function of x, the derived function, hitherto denoted by $D_x y$, is often represented by dy/dx, and the second derived function $D_x^2 y$ by d^2y/dx^2.

For example, if $y = 2x^3 + 3x^2 - 7x + 11$, we can write

$$\frac{dy}{dx} = 6x^2 + 6x - 7 \quad \text{and} \quad \frac{d^2y}{dx^2} = 12x + 6.$$

There is some disadvantage in the notation, since it tends to suggest that the derivatives are fractions rather than functions. There are also advantages which will become evident when the notation is used; for example, it makes the chain rule,

$$\frac{dy}{dx} = \frac{dy}{du} \cdot \frac{du}{dx},$$

very easy to remember.

If D_x is thought of as an *operator*, or an instruction to carry out the process of differentiation, then the corresponding operator is d/dx; similarly D_x^2 ,the instruction to differentiate twice, becomes $d^2/(dx^2)$. Thus

$$\frac{d}{dx}(2x^3 + 3x^2 - 7x + 11) = 6x^2 + 6x - 7,$$

and

$$\frac{d^2}{dx^2}(2x^3 + 3x^2 - 7x + 11) = 12x + 6.$$

The reader is advised to become familiar with this notation originally due to *Leibniz*, as he is likely to meet it frequently.

EXERCISE 13.3

1. A particle starts from rest and travels in a straight line with constant acceleration a ft/sec². After travelling s feet its velocity is v ft/sec. Express v in terms of a and s, and also v in terms of a and t, where t seconds is the time taken to travel s feet.

During the fifth second of the motion the particle travels 6 ft. Find how far it has travelled when its velocity is 8 ft/sec.

2. A particle starts from rest and has an acceleration of $(8 - 2t)$ cm/sec², where t seconds is the time measured from the instant the particle begins to move. Calculate (i) the velocity after 2 sec; (ii) the distance travelled during the third second.

3. A particle moves along the x-axis and its position after t seconds is given by

$$x = 27 - 36t + 12t^2 - t^3.$$

Show that the particle moves in the positive direction for only 4 sec, and find the acceleration at the two times when the velocity is zero.

4. Find $D_y x$ when $x = 3y - y^2 - 3$. Hence find the gradient of the curve whose equation is $y^2 + x - 3y + 3 = 0$ at the point $(-1, 2)$.

5. Solve the differential equation

$$\frac{dy}{dx}(2y + 3) = 1,$$

given that $y = 1$ when $x = 3$.

6. If $x^2 + 2xy + 3y^2 = 1$, find dy/dx and prove that

$$(x + 3y)^3 \frac{d^2y}{dx^2} + 2 = 0.$$

7. If $x = t + 1/t$, $y = t - 1/t$ find dy/dx and d^2y/dx^2 in terms of t.

8. If $x = 3t + t^3$, $y = 3 - t^{5/2}$, express dy/dx and d^2y/dx^2 in terms of t. Find the values of x when $d^2y/dx^2 = 0$.

9. A curve, whose equation is $y = f(x)$, is such that

$$\frac{d^2y}{dx^2} = ax + b.$$

The curve has a minimum at $(2, -19)$ and a maximum when $x = -3$. Prove that $a = 2b$ and $a > 0$.

If the curve passes through $(3, -21/2)$, find its equation.

10. If $dy/dx = p$, show that $d^2y/dx^2 = p(dp/dy)$. If $d^2y/dx^2 + 4y = 10$, and $dy/dx = 0$ when $y = 0$, find the value of dy/dx when $y = \frac{5}{3}$.

CHAPTER 14

INTEGRATION AND APPLICATIONS

An important limit

Let $y = f(x)$ be a function of x with domain $a \leqslant x \leqslant b$ and range $L \leqslant y \leqslant U$, where L and U are finite.

Divide $a \leqslant x \leqslant b$ at $x = x_1, x = x_2, \ldots x = x_{n-1}$ in the order $a < x_1 < x_2 < x_3 < \ldots < x_{n-1} < b$, thereby creating n sub-intervals $[x_{r-1}, x_r]$.

Suppose that, when x lies in the rth sub-interval $[x_{r-1}, x_r]$,

$$L_r \leqslant f(x) \leqslant U_r.$$

Consider the two sums:

$$S_n = U_1(x_1 - a) + U_2(x_2 - x_1) + \ldots + U_n(b - x_{n-1}),$$
$$s_n = L_1(x_1 - a) + L_2(x_2 - x_1) + \ldots + L_n(b - x_{n-1}).$$

Then

$$L(b - a) \leqslant s_n \leqslant S_n \leqslant U(b - a).$$

Now let $n \to \infty$ in such a way that $(x_r - x_{r-1}) \to 0$ for every sub-interval. If s_n and S_n both tend to the same common limit I, this common limit is called the *definite integral* of $f(x)$ over the interval $a \leqslant x \leqslant b$, and we write

$$I = \int_a^b f(x)\,\mathrm{d}x.$$

The sign $\int$ is an elongated S, for sum, and the dx originates from dy/dx. $\int_a^b f(x)\,\mathrm{d}x$ is usually read, "the integral of $f(x)$ from a to b".

The basic idea of an integral is found in the works of *Archimedes* (287–212 B.C.), but the definition given here is due to *Riemann* (1826–66). A more general definition of an integral was given by *Lebesgue* in 1902.

An example of an integral

Let $y = x^2$ when $0 \leqslant x \leqslant a$, and divide the interval $[0, a]$ into n equal sub-intervals Δx, so that $n \,.\, (\Delta x) = a$.

With the notation of the previous section,

$$L = 0, \quad U = a^2, \quad L_r = \{(r-1)\,.\,\Delta x\}^2, \quad U_r = \{r\,.\,\Delta x\}^2,$$

and so

$$\begin{aligned} s_n &= 0\,.\,\Delta x + \{1\,.\,\Delta x\}^2\,.\,\Delta x + \{2\,.\,\Delta x\}^2\,.\,\Delta x + \dots \\ &\qquad\qquad + \{(n-1)\,.\,\Delta x\}^2\,.\,\Delta x \\ &= (\Delta x)^3\,.\,\{1^2 + 2^2 + 3^2 + \dots + (n-1)^2\} \\ &= (\Delta x)^3\,.\,\tfrac{1}{6}n(n-1)(2n-1) \\ &= \tfrac{1}{6}\,\{n\,.\,\Delta x\}\,.\,\{n\,.\,\Delta x - \Delta x\}\,.\,\{2n\,.\,\Delta x - \Delta x\} \\ &= \tfrac{1}{6}\,a(a - \Delta x)(2a - \Delta x). \end{aligned}$$

Hence

$$\lim_{n\to\infty} s_n = \tfrac{1}{3}a^3.$$

Again,

$$\begin{aligned} S_n &= \{1\,.\,\Delta x\}^2\,.\,\Delta x + \{2\,.\,\Delta x\}^2\,.\,\Delta x + \{3\,.\,\Delta x\}^2\,.\,\Delta x + \dots \\ &\qquad\qquad + \{n\,.\,\Delta x\}^2\,.\,\Delta x \\ &= (\Delta x)^3\,.\,\{1^2 + 2^2 + 3^2 + \dots + n^2\} \\ &= (\Delta x)^3\,.\,\tfrac{1}{6}n(n+1)(2n+1) \\ &= \tfrac{1}{6}\,\{n\,.\,\Delta x\}\,\{n\,.\,\Delta x + \Delta x\}\,\{2n\,.\,\Delta x + \Delta x\} \\ &= \tfrac{1}{6}\,a(a + \Delta x)(2a + \Delta x), \end{aligned}$$

and so

$$\lim_{n\to\infty} S_n = \tfrac{1}{3}a^3.$$

Therefore s_n and S_n both tend to the same limit $\tfrac{1}{3}a^3$, and we write

$$I = \int_0^a x^2\,\mathrm{d}x = \tfrac{1}{3}a^3.$$

The method of selecting the sub-intervals Δx in this example is very much more restricted than in the general definition of I, but it can be shown that the result does not depend on the particular method of subdivision.

Ex. 1. Using the method of the previous example, evaluate (i) $\int^a (2x + 3)\mathrm{d}x$; (ii) $\int_1^2 (3x^2 + 2)\mathrm{d}x$.

Area by integration

Figure 14.1 represents part of the graph of $y = x^2$. Let $OA = a$, and divide OA into n parts each of length Δx, so that $n \,.\, \Delta x = a$.

Suppose MN is the rth part, so that $OM = (r - 1) . \Delta x$, $ON = r \,.\, \Delta x$, $MP = \{(r - 1) \,.\, \Delta x\}^2$, $NQ = \{r \,.\, \Delta x\}^2$.

The sum of the areas of all the inside rectangles like $MNRP$ is

$$s_n = \sum_{r=1}^{n} \{(r - 1) \,.\, \Delta x\}^2 \,.\, \Delta x,$$

where s_n has the same meaning as in the previous section, so that

$$\lim_{n \to \infty} s_n = \tfrac{1}{3}a^3.$$

The sum of all the outside rectangles like $MNQS$ is

$$S_n = \sum_{r=1}^{n} \{r \,.\, \Delta x\}^2 \,.\, \Delta x,$$

and again S_n has the same meaning as in the previous section, so that $\lim\limits_{n \to \infty} S_n = \tfrac{1}{3}a^3$.

In Fig. 14.1 we can think of the "area" enclosed by OA, AB and

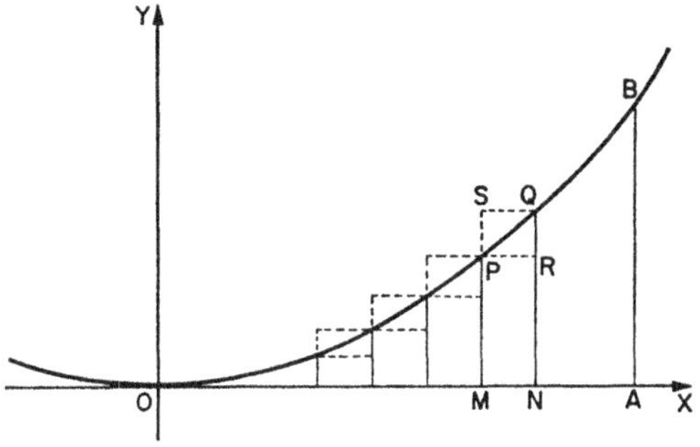

Fig. 14.1

the curve $OPQB$ as something lying between s_n and S_n for all values of n, and so we define this area to be

$$\int_0^a x^2\,dx.$$

The limiting process of evaluating $\int_a^b f(x)\,dx$ is called *integration* from first principles.

Ex. 2. Find, by integration, the area given by

$$0 \leqslant y \leqslant 2x, \quad 0 \leqslant x \leqslant a.$$

Area beneath a curve

In Fig. 14.2 the curve CD represents the graph of $y = f(x)$ between $x = a$, $x = b$.

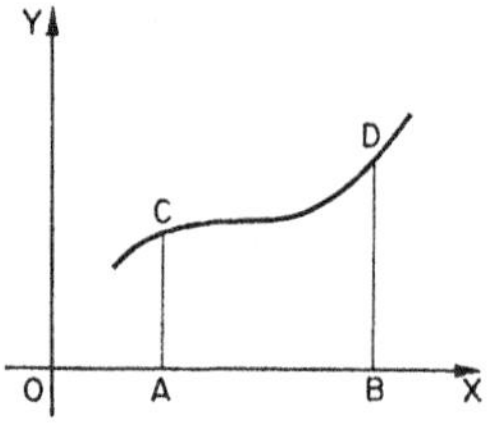

Fig. 14.2

If $OA = a$, $OB = b$, we define the area bounded by AC, BD, AB and the curve CD to be

$$\int_a^b f(x)\,dx.$$

It will be clear from the method of evaluating an integral that its value depends only on a, b and the form of the function f; it does not depend on x. In fact

$$\int_a^b f(x)\,dx = \int_a^b f(u)\,du.$$

A function defined as a definite integral

Now that we have the analogy of an area to give a concrete view of a definite integral, it is clear that $\int_a^x f(u)\,du$ is a function of x.

We shall assume that $\int_a^c f(u)\,du + \int_c^b f(u)\,du = \int_a^b f(u)\,du$, a result that appears "obvious", but for which a rigorous proof involves more difficult theorems on limits than we have dealt with in this book.

Suppose

$$z = \int_a^x f(u)\,du,$$

so that

$$z + \Delta z = \int_a^{x+\Delta x} f(u)\,du,$$

$$= \int_a^x f(u)\,du + \int_x^{x+\Delta x} f(u)\,du.$$

Hence

$$\Delta z = \int_x^{x+\Delta x} f(u)\,du.$$

When u lies in the interval $[x, x + \Delta x]$, suppose

$$L \leqslant f(u) \leqslant U;$$

then

$$L \,.\, \Delta x \leqslant \Delta z \leqslant U \,.\, \Delta x$$

so that

$$L \leqslant \frac{\Delta z}{\Delta x} \leqslant U.$$

As $\Delta x \to 0$, both L and U approach the value $f(x)$. Hence

$$\frac{dz}{dx} = f(x),$$

that is

$$\frac{d}{dx}\left\{\int_a^x f(u)\,du\right\} = f(x).$$

The fundamental theorem

Figure 14.3 shows part of the graph of $y = f(x)$, between $x = a$ and $x = b$.

Suppose $OA = a$, $OB = b$ and let $OM = x$. Let A denote the

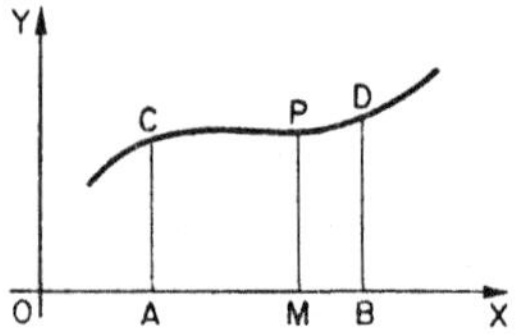

Fig. 14.3

area between the curve, the x-axis, the fixed ordinate AC and the variable ordinate MP. Then

$$A = \int_a^x f(u)\,du,$$

and so

$$\frac{dA}{dx} = f(x) = y.$$

The area function A is therefore found by solving the differential equation $dA/dx = f(x)$.

Suppose $F(x)$ is a function such that

$$\frac{d}{dx}\{F(x)\} = f(x).$$

Then

$$A = F(x) + C,$$

where C is a constant.

Now $A = 0$ when $x = a$, so that $C = -F(a)$. The area $ABDC$ is given by $x = b$, and is

$$F(b) + C = F(b) - F(a).$$

We now have a much easier method of evaluating the definite integral $\int_a^b f(x)\,dx$. First find a function $F(x)$ such that

$$\frac{d}{dx} F(x) = f(x);$$

then $\int_a^b f(x)\,dx = F(b) - F(a)$. This is known as *the fundamental theorem of the integral calculus.*

Square bracket notation

An alternative commonly used for $F(b) - F(a)$ is $[F(x)]_a^b$, and so

$$\int_a^b f(x)\,dx = [F(x)]_a^b = F(b) - F(a).$$

The function $F(x)$, found from $f(x)$ by "anti-differentiation", is called the *indefinite integral* of $f(x)$, and is written $\int f(x)\,dx$.

The word *indefinite* is used since the result of anti-differentiation is a set of functions differing only by an added constant. For example, $\int x^2\,dx = \frac{1}{3}x^3 + C$, where C is a constant.

Since $[F(x) + C]_a^b = [F(x)]_a^b$, the constant C is omitted in the evaluation of a definite integral, but it must be included in the indefinite integral.

EXAMPLE. *Evaluate* $\displaystyle\int_{-1}^{2} \left(x^2 - 2 + \frac{1}{x^3}\right) dx.$

$$D\{F(x)\} = x^2 - 2 + \frac{1}{x^3}$$

$$\Rightarrow F(x) = \frac{1}{3}x^3 - 2x - \frac{1}{2x^2} + C.$$

Hence

$$\int_{-1}^{2}\left(x^2 - 2 + \frac{1}{x^3}\right) dx = \left[\frac{1}{3}x^3 - 2x - \frac{1}{2x^2}\right]_{-1}^{2}$$

$$= \left(\tfrac{8}{3} - 4 - \tfrac{1}{8}\right) - \left(-\tfrac{1}{3} + 2 - \tfrac{1}{2}\right)$$

$$= -2\tfrac{5}{8}.$$

EXAMPLE. *Evaluate* $\displaystyle\int_2^3 \frac{1}{\sqrt{(3x-5)}}\,dx.$

$$\frac{1}{\sqrt{(3x-5)}} = (3x-5)^{-1/2}.$$

Note that the function which, when differentiated, will have this form must include $(3x - 5)^{1/2}$, i.e. the index must increase by 1.

Now

$$D(3x-5)^{1/2} = \tfrac{1}{2}(3x-5)^{-1/2}\,.\,3$$

$$= \tfrac{3}{2}(3x-5)^{-1/2}$$

$$\Rightarrow \int(3x-5)^{-1/2}\,dx = (3x-5)^{1/2}/\tfrac{3}{2}$$

$$\Rightarrow \int_2^3 \frac{1}{\sqrt{(3x-5)}}\,dx = \Big[\tfrac{2}{3}(3x-5)^{1/2}\Big]_2^3$$

$$= \tfrac{2}{3}(4^{1/2} - 1^{1/2}) = \tfrac{2}{3}.$$

This suggests that a method of integration is to look for a likely function, differentiate it, and then complete the integration as the inverse process.

Ex. 3. Find (i) $\int\left(3x^2 - \frac{1}{x^2}\right)dx$; (ii) $\int(2x+3)^3dx$; (iii) $\int\frac{3}{(2x-1)^2}\,dx$;

Ex. 4. Evaluate (i) $\int_{-1}^{1}\left(2x^3 - 3x + 1 - \frac{1}{x^2}\right)dx$; (ii) $\int_0^1(3x-1)^4dx$;

(iii) $\int_0^{1\frac{1}{2}}\frac{1}{(2x+1)^3}\,dx$; (iv) $\int_1^{11/4}\sqrt[3]{(4x-3)}dx$.

Applications of integration

Definite integrals can be applied to find not only areas, but also volumes, considered as the limit sum of elements. We first find an expression for the area or volume of the element, and then express the limit sum of the elements as a definite integral.

EXAMPLE. *Find the area bounded by*

$$y = x^3 + 3,\quad y = 0,\quad x = -1 \text{ and } x = 2.$$

The area of any element is approximately $y\,.\,\Delta x$. [A more exact statement is that it lies between $y \,.\, \Delta x$ and $(y + \Delta y) \,.\, \Delta x$; see Fig. 14.4.]

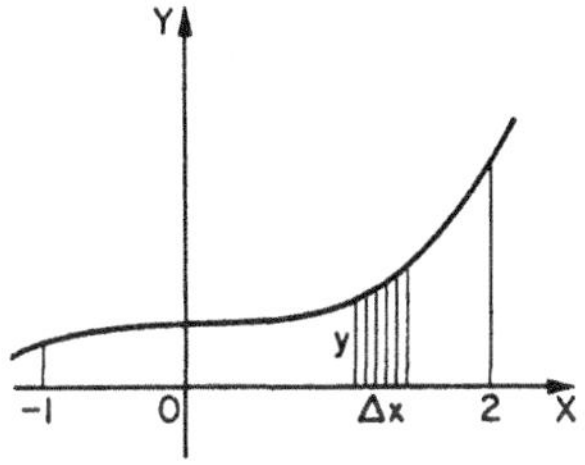

Fig. 14.4

$$\text{The area is } \lim_{\Delta x \to 0} \sum_{x=-1}^{x=2} (x^3 + 3) \,.\, \Delta x$$

$$= \int_{-1}^{2} (x^3 + 3)\, dx$$

$$= \left[\tfrac{1}{4}x^4 + 3x\right]_{-1}^{2}$$

$$= (4 + 6) - (\tfrac{1}{4} - 3)$$

$$= 12\tfrac{3}{4}.$$

In each of the following Exercises, first sketch the area to be found.

Ex. 5. Find the area between $y = 0$, $y = 1/x^2$, $x = 1$, $x = 4$.

Ex. 6. Find the area enclosed between $y = x^2 - 2x - 3$ and the line $y = 0$. Why is the result negative?

Ex. 7. Find the area between $y = x^3 + x + 2$ and $y = 0$, (i) from $x = -2$ to $x = -1$; (ii) from $x = -1$ to $x = 2$. Find the numerical sum of the areas. Compare the result with $\int_{-2}^{2}(x^3 + x + 2)dx$.

Ex. 8. Find the area bounded by $y = x^2$, $x = 0$, $y = 1$ and $y = 4$. [Use the element $x \,.\, \Delta y$.]

EXAMPLE. *Find the area enclosed between the curve*

$$x = y^2 - y - 3 \text{ and the line } x = -1.$$

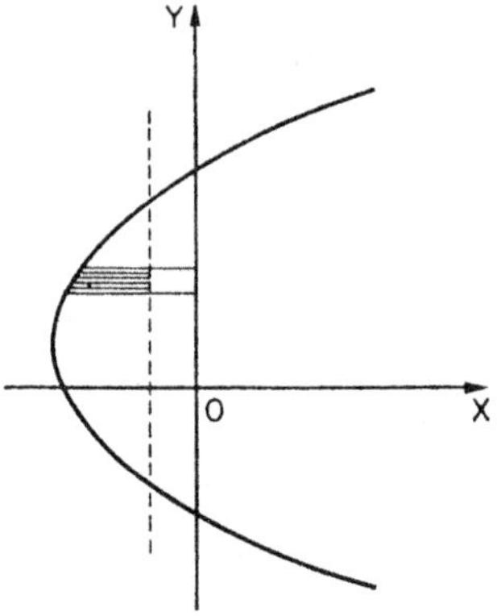

Fig. 14.5

In Fig. 14.5 the curve $x = y^2 - y - 3$ is cut by the line $x = -1$ where

$$y^2 - y - 3 = -1,$$

$$y^2 - y - 2 = 0,$$

$$(y + 1)(y - 2) = 0,$$

i.e. where $y = -1$ or 2.

The area of the shaded element is approximately

$$\{-1 - (y^2 - y - 3)\} . \Delta y.$$

Hence the area is

$$\lim_{\Delta y \to 0} \sum_{y=-1}^{y=2} (2 + y - y^2) . \Delta y$$

$$= \int_{-1}^{2} (2 + y - y^2) \, dy = \left[2y + \tfrac{1}{2}y^2 - \tfrac{1}{3}y^3\right]_{-1}^{2}$$

$$= (3\tfrac{1}{3}) - (-1\tfrac{1}{6}) = 4\tfrac{1}{2}.$$

EXERCISE 14.1

1. Find (i) $\int(3x^3 - 2x + 1)\mathrm{d}x$; (ii) $\int(2x - 5)^5\mathrm{d}x$; (iii) $\int \frac{1}{(3x+1)^4}\mathrm{d}x$;

(iv) $\int\sqrt{(x+6)}\mathrm{d}x$; (v) $\int(2x+3)^{-1/3}\mathrm{d}x$; (vi) $\int\left(\sqrt{x} - \frac{1}{\sqrt{x}}\right)\mathrm{d}x$.

2. Evaluate (i) $\int_{-1}^{1}(3x^2 + 2x + 1)\mathrm{d}x$; (ii) $\int_{1}^{2}\left(5x - \frac{1}{x^2}\right)\mathrm{d}x$;

(iii) $\int_{2}^{4}\left(x + \frac{1}{x^3}\right)\mathrm{d}x$; (iv) $\int_{-1}^{1}\sqrt{(x+2)}\mathrm{d}x$; (v) $\int_{-1}^{0}(3 - 2x)^5\mathrm{d}x$;

(vi) $\int_{0}^{6}\frac{1}{\sqrt{(4x+3)}^3}\mathrm{d}x$.

3. There is something about each of the following integrals which makes evaluation impossible with the knowledge gained so far. Say what it is.

(i) $\int_{0}^{1}\sqrt{(x-1)}\mathrm{d}x$; (ii) $\int_{-1}^{1}\frac{\mathrm{d}x}{x^3}$; (iii) $\int_{0}^{1}\frac{\mathrm{d}x}{x}$; (iv) $\int_{0}^{1}\sqrt{(1+x^2)}\mathrm{d}x$.

4. By the use of symmetry, try to reduce each of the following definite integrals to integrals over the interval $0 \leqslant x \leqslant 1$.

(i) $\int_{-1}^{1}(x^2+1)\mathrm{d}x$; (ii) $\int_{-1}^{1}(x^3+x)\mathrm{d}x$; (iii) $\int_{-1}^{1}(x^2+x)\mathrm{d}x$.

5. Find the area between (i) $y = 2x^3$, $y = 0$, $x = 1$, $x = 2$; (ii) $y = 2x^3$, $x = 0$, $y = 2$, $y = 16$.

6. Find the area cut off between $y = 0$ and (i) $y = x^2 - 4x + 3$; (ii) $y = 1 - x - 6x^2$, and between $x = 0$ and (iii) $2x = y^2 - 1$, (iv) $x = 2 - 5y - 3y^2$.

7. Find the area of the region $\{(x, y)|x^2 - x - 1 \leqslant y \leqslant -1\}$.

8. If A is $\{(x, y)|y > x\}$ and B is $\{(x, y)|\ y < 2 - x^2\}$ find the area of $A \cap B$.

9. Find the area enclosed between $x = 1$ and $x = y^2 - 5y - 5$.

10. Find the area for which $x > 0$, $y > 0$, enclosed between $y = 1/x^2$, $y = 4$, $x = 3$.

11. Find the area enclosed between $y^2 = 3x$ and $x = 3$.

12. Find the area enclosed between $y^2 = 4x$ and $y^2 = 4(4 - x)$.

13. Find the area enclosed between $y = x^2 - 5x$ and $y = 4 - 2x$.

14. Find the area between $y = 1/x^2$, $y = 0$, $x = 1$ and $x = X$. Has this area a limiting value as $X \to \infty$? Repeat for the interval $x = x_1$ to 1, where $0 < x_1 \leqslant 1$. Has this area a limiting value as $x_1 \to 0$?

15. Find a numerical value for the sum of the areas cut off between the curve $y = x^2 - 1$ and the x-axis for the interval $-2 \leqslant x \leqslant 2$.

16. Find the area enclosed between part of the curve $y = (1 - x)^2(4 - x)$ and the x-axis.

Volume of a solid of revolution

Figure 14.6 represents a portion of a curve whose equation is $y = f(x)$. The area bounded by the arc $CPQD$, the x-axis and the ordinates AC, BD is revolved through four right-angles about the axis OX.

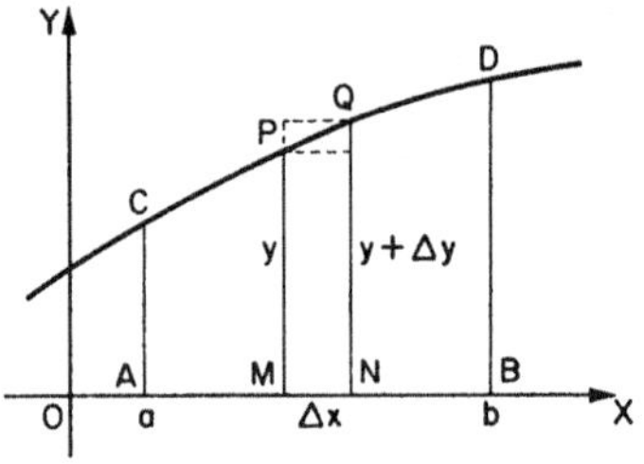

Fig. 14.6

$PMNQ$ is an element of width Δx; $PM = y$ and $QN = y + \Delta y$. The element, ΔV, of the solid of revolution, due to the element $PMNQ$ is such that

$$\pi y^2 \,.\, \Delta x < \Delta V < \pi(y + \Delta y)^2 \,.\, \Delta x.$$

If both $\lim\limits_{\Delta x \to 0} \sum\limits_{x=a}^{x=b} \pi y^2 \,.\, \Delta x$ and $\lim\limits_{\Delta x \to 0} \sum\limits_{x=a}^{x=b} \pi(y + \Delta y)^2 \,.\, \Delta x$ exist and are equal, their value is $\int_a^b \pi y^2 \, dx$.

Hence the volume formed by rotating the region

$$\{(x, y) | a \leqslant x \leqslant b, \quad 0 \leqslant y \leqslant f(x)\}$$

about OX is defined as

$$V = \int_a^b \pi y^2 \, \mathbf{d}x.$$

Similarly, the volume of a solid of revolution formed by a complete revolution about OY is $\int_c^d \pi x^2 \, dy$.

EXAMPLE. *Find the formula for the volume of a cone of base radius r and height h.*

Consider a section of the cone perpendicular to the axis, at a distance y from the vertex. (See Fig. 14.7.) This section is a circle of radius x, where, by similar triangles,

$$\frac{x}{r} = \frac{y}{h}, \quad \text{so } x = \frac{r}{h}y.$$

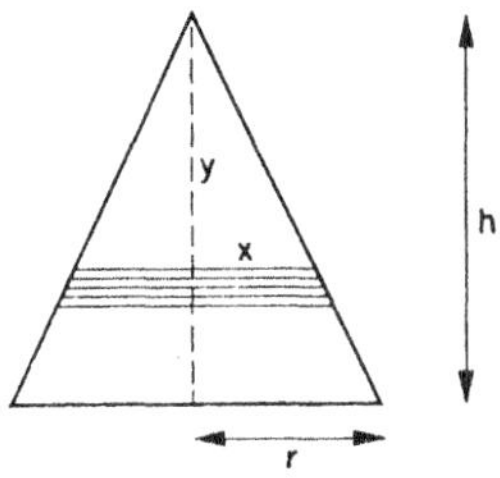

Fig. 14.7

An element of thickness Δy is a disc, whose volume $\Delta V \approx \pi x^2 \,.\, \Delta y$, and so the volume of the cone is

$$\lim_{\Delta x \to 0} \sum_{y=0}^{y=h} \pi x^2 \,.\, \Delta y$$

$$= \int_0^h \pi \frac{r^2}{h^2} y^2 \,.\, \mathrm{d}y = \frac{\pi r^2}{3h^2} \left[y^3\right]_0^h$$

$$= \tfrac{1}{3}\pi r^2 h.$$

Ex. 9. Prove the formula for the volume of a sphere of radius r.

EXAMPLE. *Find the volume of the region*

$$\{(x, y, z) | a \leqslant z \leqslant b, \quad cz \geqslant x^2 + y^2.\}$$

Figure 14.8 shows the section of $cz = x^2 + y^2$ by the plane $x = 0$, i.e. the curve $cz = y^2$ in $x = 0$. Rotation of this curve about the z-axis gives the surface $cz = x^2 + y^2$, and $cz > x^2 + y^2$ *inside* this surface. The required volume is enclosed within this surface and the planes $z = a$, $z = b$.

An element at height z and with thickness Δz has a volume, $\Delta V \approx \pi y^2 . \Delta z$. Hence the volume of the solid of revolution is

$$\lim_{\Delta z \to 0} \sum_{z=a}^{z=b} \pi . cz . \Delta z = \pi c \int_a^b z \, dz$$
$$= \tfrac{1}{2}\pi c(b^2 - a^2).$$

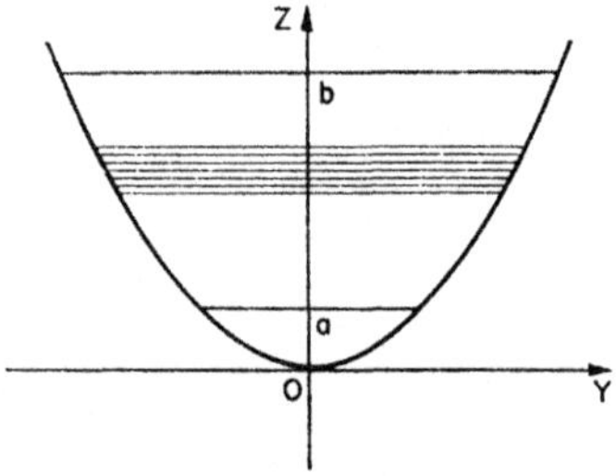

Fig. 14.8

Ex. 10. Find the volume of the solid $\{(x, y, z) | p \leqslant x \leqslant q, x^2 \geqslant y^2 + z^2\}$. Describe the solid.

Ex. 11. Figure 14.9 shows a section of a solid of revolution, formed by revolving the area beneath a portion of the curve $y = f(x)$ about OX. V is the volume of the solid from a fixed value $x = a$ to a variable value x. The corresponding element of volume has a thickness Δx. Show that

$$D_x V = \pi\{f(x)\}^2.$$

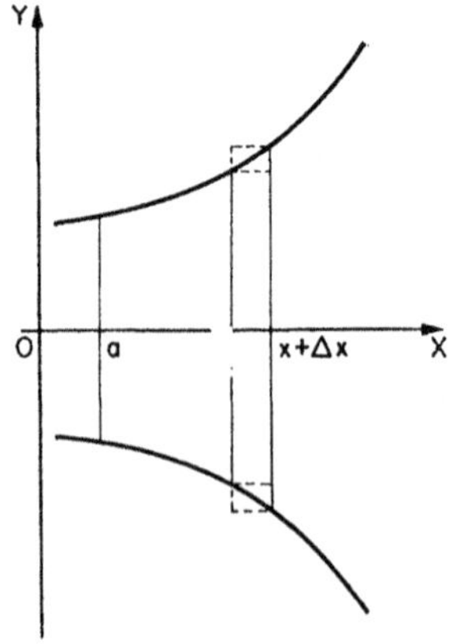

Fig. 14.9

EXERCISE 14.2

1. Find the volume formed by the rotation about the x-axis of the part of the curve $x^2 + y^2 = 9$ between $x = 1$ and $x = 2$.

2. The line $x - y = 4$ is rotated about the y-axis. Find the volume of the frustum of the cone so formed between $y = 0$ and $y = 3$.

3. Sketch the curve $ax^2 = y^3$, where $a > 0$. If this curve is rotated about the y-axis find the volume enclosed between $y = 0$ and $y = 2a$.

4. A vase is formed by rotating about the axis $x = 0$ the part of the curve $x^2 = y(y - 5)^2$ that lies between $y = 0$ and $y = 4$. Find the volume of the vase.

5. A bowl has the shape of a spherical cup. It is part of a sphere of radius a cm and has a depth h cm in the middle. Find the volume of water it would hold.

6. Find the volume formed by rotating the ellipse $4x^2 + 9y^2 = 36$, (i) about the x-axis, (ii) about the y-axis.

7. Find the volume formed by rotating about the x-axis the area enclosed between $y = x^2 - 6x + 5$ and the x-axis.

8. The curve $x = y(y - 2)(y - 3)$ is rotated about the y-axis. Find (i) the numerical value of the area enclosed between the curve and the y-axis, (ii) the volume formed by rotating this area about the y-axis.

9. Find the volume of the region $\{(x, y, z)|y \leqslant 4, 2y \geqslant x^2 + z^2\}$.

10. Find the volume of the region

$$\{(x, y, z)|0 \leqslant z \leqslant 40, (z + 20)^2 \geqslant 4(x^2 + y^2)\}.$$

Mean values

The *mean value* of the set of n numbers $y_1, y_2, y_3, \ldots, y_n$ is defined to be

$$\frac{y_1 + y_2 + y_3 + \ldots + y_n}{n}.$$

If y is a function of x, say $y = f(x)$, the mean value of y over the interval $a \leqslant x \leqslant b$ would appear to be the height of the rectangle with base $(b - a)$ which has the same area as the area beneath the curve $y = f(x)$, that is:

$$\frac{1}{b - a}\int_a^b f(x)\, dx.$$

We can arrive at this result in another way. Suppose the interval $a \leqslant x \leqslant b$ is divided into n sub-intervals, each equal to Δx, so that $n\,.\,\Delta x = b - a$. Let the values of y for $x = a, a + \Delta x, a + 2\,.\,\Delta x, \ldots, a + (n-1)\,\Delta x$ be $y_1, y_2, y_3, \ldots, y_n$. The mean value of the n values of y is

$$\frac{1}{n}\sum_{r=1}^{n} y_r = \frac{\sum\limits_{r=1}^{n} y_r\,.\,\Delta x}{n\,.\,\Delta x} = \frac{\sum\limits_{r=1}^{n} y_r\,.\,\Delta x}{b-a}.$$

If the number of sub-intervals is increased so that $\Delta x \to 0$, the mean value approaches the limiting value

$$\frac{1}{b-a}\int_a^b f(x)\,dx.$$

Another mean value frequently needed in statistics is the *root mean square*; for the n numbers $y_1, y_2, y_3, \ldots, y_n$, this is defined to be

$$\sqrt{\left\{\frac{1}{n}(y_1^2 + y_2^2 + y_3^2 + \ldots + y_n^2)\right\}}.$$

For the function $y = f(x)$ over the interval $a \leqslant x \leqslant b$ the root mean square is

$$\sqrt{\left\{\frac{1}{b-a}\int_a^b y^2\,dx\right\}}.$$

EXAMPLE. *Find the mean value of positive ordinates of* $y^2 = 4ax$ *from* $x = 0$ *to* $x = a$.

The required mean value is

$$\frac{1}{a}\int_0^a \sqrt{(4ax)}\,dx = \frac{2\sqrt{a}}{a}\left[\tfrac{2}{3}x^{3/2}\right]_0^a$$
$$= \tfrac{4}{3}a.$$

EXAMPLE. *Find the mean value and the root mean square of*

$$y = x(x-1)(x-2) \quad \textit{for} \quad 0 \leqslant x \leqslant 2.$$

The mean value is

$$\tfrac{1}{2}\int_0^2 (x^3 - 3x^2 + 2x)\,dx = \tfrac{1}{2}\Big[\tfrac{1}{4}x^4 - x^3 + x^2\Big]_0^2 = 0.$$

The root mean square is

$$\sqrt{\{\tfrac{1}{2}\int_0^2 (x^3 - 3x^2 + 2x)^2\,dx\}}$$

$$= \sqrt{\{\tfrac{1}{2}\int_0^2 (x^6 - 6x^5 + 13x^4 - 12x^3 + 4x^2)\,dx\}}$$

$$= \sqrt{\{\tfrac{1}{2}\Big[\tfrac{1}{7}x^7 - x^6 + \tfrac{13}{5}x^5 - 3x^4 + \tfrac{4}{3}x^3\Big]_0^2\}}$$

$$= \sqrt{(\tfrac{8}{105})} \approx 0{\cdot}276.$$

Centre of mass

Suppose a set of particles of masses $m_1, m_2, m_3, \ldots$ is located at the points $(x_1, y_1), (x_2, y_2), (x_3, y_3), \ldots$. The *centre of mass* of the particles is the point $(\bar{x}, \bar{y})$ where

$$\bar{x} = \frac{\sum m_r x_r}{\sum m_r}, \quad \bar{y} = \frac{\sum m_r y_r}{\sum m_r}.$$

An expression like $\sum m_r x_r / \sum m_r$ is called a *weighted* mean value, and is the same as an ordinary mean value when

$$m_1 = m_2 = m_3 = \ldots.$$

In the case of a lamina or solid (the definition is readily extended to three dimensions) of continuous structure, a summation such as $\sum m_r x_r$ over the whole area or volume becomes a definite integral. For a region $a \leqslant x \leqslant b$, $0 \leqslant y \leqslant f(x)$ the results are as follows:

Lamina: $$\bar{x} = \frac{\int_a^b xy\,dx}{\int_a^b y\,dx}, \quad \bar{y} = \frac{\int_a^b \frac{1}{2}y^2\,dx}{\int_a^b y\,dx}.$$

Solid of revolution: $\bar{x} = \dfrac{\int_a^b \pi x y^2 \, dx}{\int_a^b \pi y^2 \, dx}, \qquad \bar{y} = 0,$

the axis of revolution being the x-axis.

The process will be illustrated in the following Example.

EXAMPLE. *Find the centre of mass of a uniform triangular lamina* OBA *with* $\widehat{B} = 90°$, $OB = b$, $BA = a$. *Also find the centre of mass of the cone formed by revolving the* ΔOBA *about the side* OB.

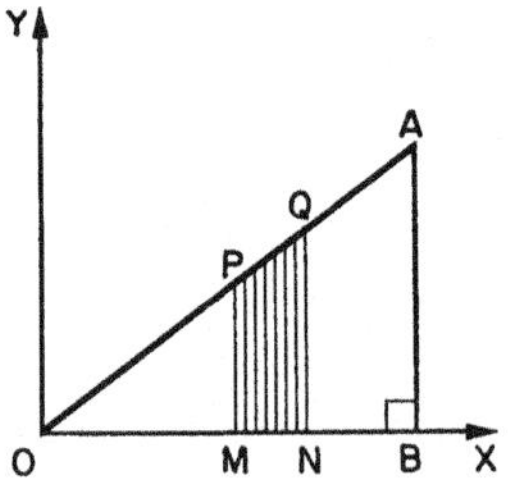

Fig. 14.10

Let OBA be the triangle, as in Fig. 14.10. If $OM = x$, $MN = \Delta x$, the element $MNQP$ has a mass $\rho y \,.\, \Delta x$ approximately, where ρ is the mass per unit area. Hence

$$\bar{x} = \frac{\lim\limits_{\Delta x \to 0} \Sigma(\rho y \,.\, \Delta x) \,.\, x}{\lim\limits_{\Delta x \to 0} \Sigma(\rho y \,.\, \Delta x)}.$$

The denominator is ρ times the area of ΔOBA. By similar triangles, $y : a = x : b$, or $y = ax/b$.

$$\bar{x} = \frac{\rho \int_0^b (ax^2/b) \, dx}{\frac{1}{2}\rho ab} = \tfrac{2}{3}b.$$

The centre of mass of the element $MNQP$ is approximately $\frac{1}{2}y$ from OX, and so

$$\bar{y} = \frac{\lim\limits_{\Delta x \to 0} \Sigma(\rho y \,.\, \Delta x) \,.\, (\frac{1}{2}y)}{\lim\limits_{\Delta x \to 0} \Sigma(\rho y \,.\, \Delta x)} = \frac{\frac{1}{2}\rho \int_0^b y^2 \, dx}{\frac{1}{2}\rho ab}$$

$$= \frac{1}{ab}\int_0^b y^2 \, dx = \frac{1}{ab} \cdot \frac{a^2}{b^2}\int_0^b x^2 \, dx = \tfrac{1}{3}a.$$

The centre of mass of the triangle is therefore $(\frac{2}{3}b, \frac{1}{3}a)$.

The cone formed by revolving ΔOBA about OB has OB as axis of symmetry, and, for the cone, $\bar{y} = 0$; also

$$\bar{x} = \frac{\lim\limits_{\Delta x \to 0} \Sigma(d\pi y^2 \,.\, \Delta x) \,.\, x}{\lim\limits_{\Delta x \to 0} \Sigma(d\pi y^2 \,.\, \Delta x)},$$

where d is the mass per unit volume; then

$$\bar{x} = \frac{\int_0^b \pi x y^2 \, dx}{\int_0^b \pi y^2 \, dx} = \frac{\int_0^b x^3 \, dx}{\int_0^b x^2 \, dx} = \tfrac{3}{4}a.$$

The centre of mass of the cone is on OB, and $\bar{x} = \frac{3}{4}a$.

Ex. 12. Find the centre of mass of the area in the first quadrant enclosed by $y^2 = 4ax$, $y = 0$, $x = b$, and also of the solid formed by revolving this area about $y = 0$.

Numerical integration

The process of evaluating a definite integral, as we have seen it so far, is carried out either by finding the limit sum of a series or by finding the indefinite integral of a function. Both these methods fail, or become extremely difficult, for many functions which have a deceptively simple appearance, such as $\int_0^1 \sqrt{(1 + x^3)} \, dx$. The representation of a definite integral by the area under a graph allows us, in many cases, to evaluate the definite integral within as narrow a limit of accuracy as we wish. Two methods will be used, both being very well suited to machine computation.

The trapezium rule

The problem is to evaluate $\int_a^b f(x)\,dx$, and we assume that it is possible to evaluate $f(x)$ when $a \leqslant x \leqslant b$.

Divide the interval $a \leqslant x \leqslant b$ into n equal sub-intervals, each of length h, so that $nh = b - a$. Let $y_1, y_2, y_3, \ldots y_{n+1}$ be the values of $f(x)$ when

$$x = a, \quad a + h, \quad a + 2h, \ldots, \quad a + nh.$$

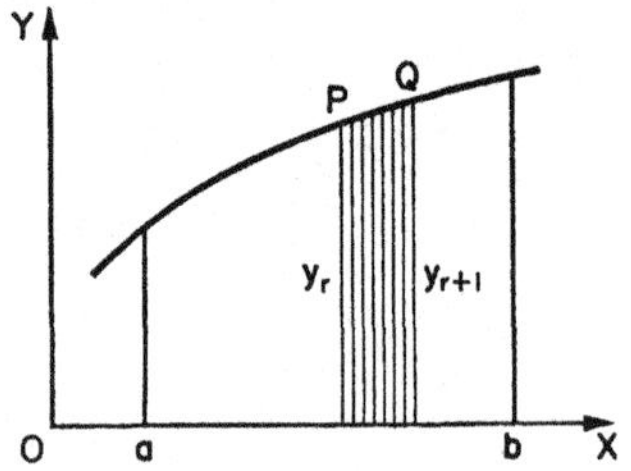

Fig. 14.11

The strip, shown in Fig. 14.11, bounded by the x-axis, the curve $y = f(x)$, and the ordinates y_r, y_{r+1}, is replaced by a trapezium by drawing the chord PQ. Then

$$\begin{aligned}\int_a^b f(x)\,dx &\approx \sum_{r=1}^{n} \tfrac{1}{2}(y_r + y_{r+1})\,.\,h \\ &= \tfrac{1}{2}h(y_1 + 2y_2 + 2y_3 + \ldots + y_{n+1}) \\ &= \tfrac{1}{2}h\{(y_1 + y_{n+1}) + 2(y_2 + y_3 + \ldots + y_n)\}.\end{aligned}$$

This rule for the approximate evaluation of $\int_a^b f(x)\,dx$ is called the *trapezium rule*. Its accuracy is increased by taking a larger number of strips, but this also increases the number of evaluations of $f(x)$. When the graph is curved in only one direction, as in Fig. 14.11, the errors, though small, are cumulative; the result obtained for $\int_a^b f(x)\,dx$ in this case would be too small.

Simpson's rule

In this method the interval $a \leqslant x \leqslant b$ is divided into an *even* number of strips, and the strips are dealt with in pairs. The curve forming the upper boundary of the two strips is replaced by a parabola of the form $y = px^2 + qx + r$, with a vertical axis of symmetry, passing through the upper ends of the ordinates forming the strip.

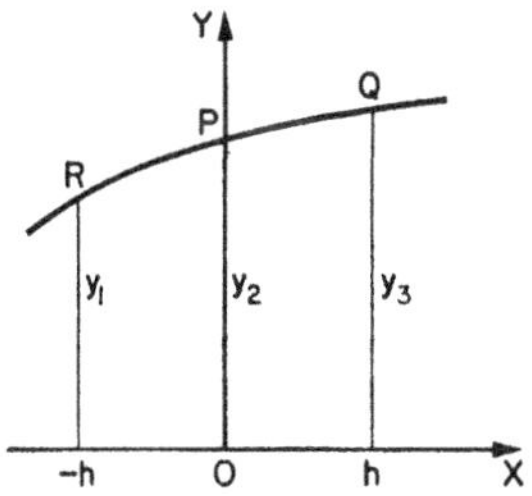

Fig. 14.12

Figure 14.12 shows two strips bounded by the ordinates y_1, y_2, y_3. The points R, P, Q are $(-h, y_1)$, $(0, y_2)$, (h, y_3). If $y = px^2 + qx + r$ passes through R, P and Q, we have

$$y_1 = ph^2 - qh + r, \quad y_2 = r, \quad y_3 = ph^2 + qh + r.$$

Solving these equations for p, q, r:

$$r = y_2, \quad q = \frac{y_3 - y_1}{2h}, \quad p = \frac{y_1 - 2y_2 + y_3}{2h^2}.$$

The area under the curve RPQ is

$$\int_{-h}^{h} (px^2 + qx + r)\,\mathrm{d}x = \left[\tfrac{1}{3}px^3 + \tfrac{1}{2}qx^2 + rx\right]_{-h}^{h}$$

$$= \tfrac{2}{3}(ph^3 + 3rh)$$

$$= \tfrac{2}{3}h\{\tfrac{1}{2}(y_1 - 2y_2 + y_3) + 3y_2\}$$

$$= \tfrac{1}{3}h(y_1 + 4y_2 + y_3).$$

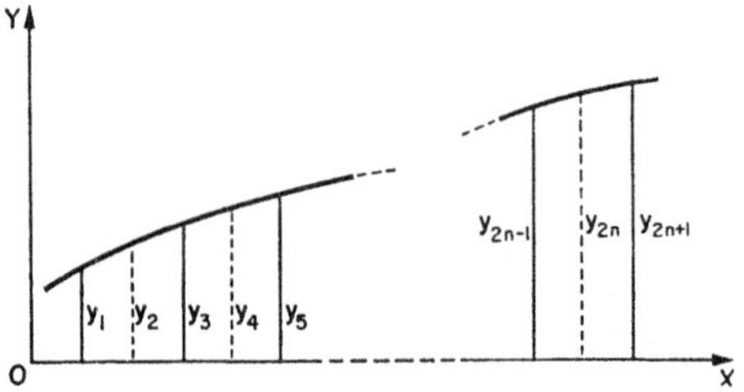

Fig. 14.13

Figure 14.13 shows the interval $a \leqslant x \leqslant b$ divided into $2n$ strips each of width h. Applying the approximation just found to each pair of strips we have:

$$\int_a^b f(x)\,dx \approx \tfrac{1}{3}h(y_1 + 4y_2 + y_3) + \tfrac{1}{3}h(y_3 + 4y_4 + y_5) + \ldots$$
$$+ \tfrac{1}{3}h(y_{2n-1} + 4y_{2n} + y_{2n+1})$$
$$= \tfrac{1}{3}h\{(y_1 + y_{2n+1}) + 4(y_2 + y_4 + \ldots + y_{2n})$$
$$+ 2(y_3 + y_5 + \ldots + y_{2n-1})\}.$$

This rule for the approximate evaluation of $\int_a^b f(x)\,dx$ is known as *Simpson's rule*.

EXAMPLE. *Evaluate* $\int_0^2 \sqrt{(1 + x^3)}\,dx$ (*i*) *by the trapezium rule,* (*ii*) *by Simpson's rule, using* 10 *strips in each case.*

Divide the interval $0 \leqslant x \leqslant 2$ into 10 sub-intervals, each of length 0·2. The following table of values is then calculated:

x	0	0·2	0·4	0·6	0·8	1·0	1·2	1·4	1·6	1·8	2·0
$1 + x^3$	1	1·008	1·064	1·216	1·512	2·0	2·728	3·744	5·096	6·832	9·0
$\sqrt{(1 + x^3)}$	1	1·004	1·032	1·103	1·230	1·414	1·651	1·935	2·257	2·613	3·0

(i) *Trapezium rule.*

$$y_1 + y_{11} = 4, \quad y_2 + y_3 + \ldots + y_{10} = 14{\cdot}239,$$
$$2(y_2 + y_3 + \ldots + y_{10}) = 28{\cdot}478.$$

$$\int_0^2 \sqrt{(1+x^3)}\,dx \approx \tfrac{1}{2}\,.\,0{\cdot}2\,.\,(4+28{\cdot}478)$$

$$= 3{\cdot}248, \quad \text{to 4 sig. fig.}$$

(ii) *Simpson's rule.*

$$y_1 + y_{11} = 4, \quad y_2 + y_4 + \ldots + y_{10} = 8{\cdot}069,$$

$$y_3 + y_5 + \ldots + y_9 = 6{\cdot}170,$$

$$4(y_2 + y_4 + \ldots + y_{10}) = 32{\cdot}276,$$

$$2(y_3 + y_5 + \ldots + y_9) = 12{\cdot}340.$$

$$\int_0^2 \sqrt{(1+x^3)}\,dx \approx \tfrac{1}{3}\,.\,0{\cdot}2\,.\,(4+32{\cdot}276+12{\cdot}340)$$

$$= 3{\cdot}241, \quad \text{to 4 sig. fig.}$$

Ex. 13. Evaluate $\int_0^1 \sqrt{x}dx$ (i) by the trapezium rule, (ii) by Simpson's rule, using 10 strips. Compare the results with that obtained by direct integration.

Ex. 14. Evaluate $\int_1^2 \log_{10}x dx$ by Simpson's rule, using 10 strips and 4-figure logarithm tables.

EXERCISE 14.3

1. Find the mean value of the ordinates of the parabola $y^2 = x + 2$ between $x = 2$ and $x = 7$.

2. Find the mean value of the velocity, v ft/sec, of a body for equal intervals of time during the first 4 sec of its motion, when $v = 8 - 3t$.

3. A volume (v cm³) of gas under a pressure of p g/cm² expands in accordance with the law $pv^{1\cdot2} = 600$. Calculate the mean value of the pressure when the volume changes from 32 cm³ to 243 cm³.

4. A sphere of radius 2 ft begins to expand so that its radius increases uniformly at the rate of 1 in/sec. Find, in ft³ the mean value of the volume (i) for equal intervals of time during the first 12 sec, (ii) for equal increases in the radius from 2 ft to 3 ft.

5. A stone falls and its distance s cm from its starting point after t sec is given by $s = 490t^2$. Find the mean value of its velocity over the first 250 cm, (i) for equal intervals of time, (ii) for equal intervals of distance. Which of the two results represents what is commonly known as "average velocity"?

6. Find the centre of mass of a uniform lamina bounded by $y^3 = a^2x$, $y = 0$, $x = b$. Find also the centre of mass of the solid formed by rotating this area about the x-axis.

7. By considering the rotation of $\{(x, y)|x \geqslant 0, x^2 + y^2 \leqslant a^2\}$ about the x-axis, find the centre of mass of a solid hemisphere of radius a.

8. A right pyramid has a square base of side a and the height of the pyramid is h. Find its centre of mass (i) when it is a uniform solid, (ii) when it is hollow and consists only of four uniform thin triangular laminas.

9. A solid is formed by rotating $\{(x, y)|0 \leqslant x \leqslant 5, y \geqslant 0, y^2 \leqslant 20x\}$ about the line $x = 5$. Find the height above the x-axis of the centre of mass of the solid.

10. A frustum of a cone has ends of radii a and b $(a > b)$, and its height is h. Find the distance of the centre of mass from the end of radius a.

11. Draw a circle of radius 3 cm and divide a diameter into 12 equal parts. Through the points of division draw chords perpendicular to the diameter. Use Simpson's rule to calculate the area of the circle, and so find an approximate value of π.

12. The relation between x and y^2 for the point (x, y) on a certain curve is shown in the table:

x	0	1	2	3	4	5	6
y^2	0	1·2	2·8	4·8	7·2	10·0	13·2

The area bounded by the curve, $y = 0$, and $x = 6$ is rotated about $y = 0$. Find the position of the centre of mass of the solid of revolution, using Simpson's rule.

13. Find $\int_0^{90} \sin x° \mathrm{d}x$ by the trapezium rule and by Simpson's rule, using 10 strips. Use 4-figure tables for $\sin x°$, and give results to 2 places of decimals.

14. Evaluate $\int_1^2 \frac{\mathrm{d}x}{x}$ by the trapezium rule and by Simpson's rule, using 10 strips and four decimal places in the working. [The result gives an approximation for $\log_e 2$; see Chapter 5, p. 75.]

15. Figures 14.14 (i) and (ii) show a portion $CPQD$ of the curve $y = f(x)$. The area bounded by the curve, AC, BD and AB, where $OA = a$ and $OB = b$, is $\int_a^b f(x)\, \mathrm{d}x$. The area is divided into n strips of unequal width in such a way that the greatest width, represented by MN, tends to zero as $n \to \infty$.

The shaded areas of (i) are shown collected together in (ii). If S_L represents the sum of the rectangles under the curve in (i), show that

$$\int_a^b f(x)\mathrm{d}x - S_L < MN\{f(b) - f(a)\}.$$

Hence show that integration, as defined on p. 202, does not depend upon a system of strips of equal width.

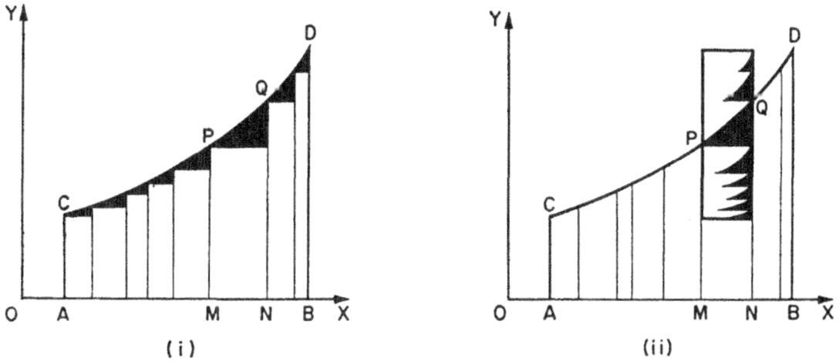

Fig. 14.14

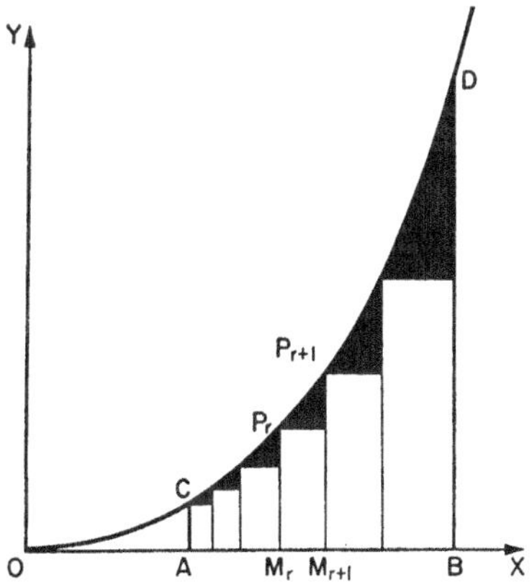

Fig. 14.15

16. Figure 14.15 shows an area $ABDC$ bounded by the curve $y = x^3$ and the lines $x = a$, $x = b$, $y = 0$. The segment AB is divided into n pieces by points $M_1, M_2, \ldots M_{n-1}$ where $OA = a$, $OM_1 = aq$, $OM_2 = aq^2, \ldots$, $OM_{n-1} = aq^{n-1}$, $OB = aq^n$.

Show that the area of the rectangle standing on M_rM_{r+1} is $a^4(q - 1)q^{4r}$. Show also that

$$0 < q - 1 < \frac{b - a}{n},$$

by considering the length of AM_1.

Deduce that the area beneath the arc CD is

$$\int_a^b x^3 dx = \frac{b^4 - a^4}{4}.$$

ANSWERS

CHAPTER 1 (pp. 1–18)

Ex. 4. 10000, 16; 101111, 47. **Ex. 5.** 0·1, 0·001, 0·011.
Ex. 6. $0{\cdot}0\dot{1}00\dot{1}$, $0{\cdot}1\dot{0}11\dot{0}$, $0{\cdot}1\dot{1}10\dot{0}$. **Ex. 7.** $\frac{3}{4}$, $2\frac{1}{4}$, $1\frac{5}{8}$.
Ex. 8. (i) 10·001, (ii) 0·01, (iii) 10·0011. **Ex. 9.** (i) 20/9, (ii) 37/33, (iii) 56/111.
Ex. 10. (i) 1, (ii) 100/11, (iii) 10011/111. **Ex. 11.** 0·3. **Ex. 13.** 17/12.
Ex. 14. $\frac{1}{2}\left(x+\frac{3}{x}\right), \ldots, \frac{1}{2}\left(x+\frac{N}{x}\right)$.

Exercise 1.1 (p. 10)

1. (i) $2\sqrt{2}$, (ii) $3\sqrt{2}$, (iii) $5\sqrt{2}$, (iv) $3{\cdot}5\sqrt{2}$. **2.** (i) $1+\sqrt{3}$, (ii) $2{\cdot}5 + 1{\cdot}5\sqrt{2}$, (iii) $4+\sqrt{15}$, (iv) $2-\sqrt{6}$. **3.** 2, 1·75, 1·732. **4.** 4. **5.** 2. **6.** −6/5, 0, 2. **7.** Any weight, to the nearest 0·5 kg, between 0 and 31·5 kg. **8.** (i) 221, 12101, 111012, 111012, 1010201; (ii) 21, 101, 258, 591. **9.** (i) 20334, (ii) 234.

Exercise 1.2 (p.17)

6. $-3 < x < 2$. **7.** $-1{\cdot}5 < x < 0{\cdot}5$. **8.** $-2 < x < 1$.
9. $x < -3$ or $x > 3$. **10.** $-5 < x < 5$. **11.** $x < -1$ or $1 < x < 3$.
12. $x < \frac{1}{3}$ or $x > \frac{1}{2}$. **13.** $x < \frac{1}{3}$ or $x > 1$. **14.** $-\frac{4}{5} < x < 2$.
15. −2, −1, 0, 1, 2, 3, 4. **16.** (i) 3, (ii) −1. **17.** (i) 5, (ii) 4.
18. (i) $x = 1$, $y = 1$ to 14; $x = 2$, $y = 1$ to 7; $x = 3$, $y = 1$ to 4; (ii) 3, 3; (iii) 1, 3; 2, 2; or 3, 1.

CHAPTER 2 (pp. 19–31)

Ex. 1. (i) Yes, (ii) yes. **Ex. 2.** $a = c$ and $b = d$, or $a = d$ and $b = c$.
Ex. 3. (i) 4, (ii) 8, (iii) 2^n. **Ex. 7.** $\{x \mid x \in A \text{ and } x \in B\}$.
Ex. 8. (i) $A = B$, (ii) $B \subseteq A$, (iii) $A \cap B = \phi$, (iv) $A \subseteq B$. **Ex. 9.** (i) **P**, **A**.

Exercise 2.1 (p. 26)

1. (i) $\{a, b, c, d, e, f, g\}$, (ii) $\{a, b\}$, (iii) $\{c, d, e, f, g\}$, (iv) $\{h\}$, (v) $\{h\}$, (vi) $\{c, d, e, f, g, h\}$, (vii) $\{a, b\}$, (viii) $\{a, b, c, d, e, f, g, h\}$. **4.** (i) $P' \cap Q'$, (ii) $P \cap Q$, (iii) $Q \cup (P \cap R)$, (iv) $P \cap (Q' \cup R')$, (v) $P' \cup (Q' \cap R')$, (vi) ϕ, (vii) $A \cap C'$.

Exercise 2.2 (p. 30)

1. $P \cup Q = Q$, $Q' \subseteq P'$. **2.** When $A \cap B = \phi$. **4.** (i) Between 18 and 25, (ii) 7 or less. **5.** Between 2 and 15. **6.** (i) 5, (ii) 7, (iii) 3. **7.** 42. **8.** 40. **9.** 10; 57; between 0 and 23. **12.** 30 per cent.

CHAPTER 3 (pp. 32–5)

EXERCISE 3.1 (p. 41)

1. (i) (10, −10), (ii) (−10, 10), (iii) (6, 2), (iv) (0, 0), (v) (3, 1), (vi) (4, 0). **3.** (i) $(2\frac{1}{2}, 5)$, (ii) $(2\frac{1}{2}, 5)$; parallelogram. **4.** $\overrightarrow{RQ} = \overrightarrow{QP}$. **5.** (2, 5), $(5\frac{1}{2}, -4)$, $(1\frac{1}{2}, 2)$; (3, 1); they are concurrent. **6.** $\frac{1}{2}R$. **7.** (3, 1). **8.** (0, 5/4). **9.** The sides and inside of the triangle given by $x = y$, $x + y = 4$, $y = 0$.

EXERCISE 3.2 (p. 45)

1. (i) (−4, 10, −2), (ii) (4, −10, 2), (iii) (0, 4, 8), (iv) (0, 0, 0), (v) (0, 2, 4), (vi) (−2/5, 3, 19/5). **2.** (i) The space between the parallel planes $x = 0$, $x = 1$, (ii) the space between the parallel planes $y = 0$, $y = \frac{1}{2}$, (iii) the space between the parallel planes $z = 0$, $z = \frac{1}{2}$, (iv) the inside of the cuboid formed by the intersection of these spaces. **3.** $(1, \frac{1}{2}, 3\frac{1}{2})$, (4, 1, 1), $(4, 3\frac{1}{2}, -\frac{1}{2})$, (1, 3, 2), $(2, 2\frac{1}{2}, \frac{1}{2})$, $(3, 1\frac{1}{2}, 2\frac{1}{2})$; each $(2\frac{1}{2}, 2, 1\frac{1}{2})$; they are concurrent. **4.** $\overrightarrow{QP} = \frac{1}{2}\overrightarrow{RQ}$. **5.** (11/3, 2, 1); $(2\frac{1}{2}, 2, 1\frac{1}{2})$; (7/3, 3, 2/3); (5/2, 3, 2/3); they are concurrent. **6.** Concurrent, at a point dividing each in the ratio 3:1. **Ex. 16.** (i) 3, (ii) 3, (iii) 4, (iv) 4, (v) 4, (vi) 2, (vii) 4, (viii) 4, (ix) 3.

EXERCISE 3.3 (p. 49)

1. (i) 4, (ii) 3, (iii) 4, (iv) 4, (v) 6, (vi) 1. **2.** 3. **3.** 4. **4.** 4, 9. **5.** 0. **6.** 1, 3, 5, 7, 9. **7.** 4, 9, 14. **8.** 2, 7. **9.** 15. **10.** No solution. **11.** (i) $c–b$ and a have no common factor, a and m have no common factor, (ii) $c–b$ and a have no commcn factor, but a and m have. **12.** (i) 0, 1, 4, (ii) 0, 1, 4, 5, 6, 9, (iii) 0, 1, 2, 3, 4, 5, 6, 7, 8, 9. **16.** (i) 1, 9, (ii) —, (iii) 2, 8, (iv) 4, 6, (v) 3, 7. **17.** Let $an - bn = mc$. Condition is $n|c$.

CHAPTER 4 (pp. 51–64)

Ex. 1. 18; (i) (1, 2), (2, 4), (3, 6); (ii) (1, 1), (2, 4). **Ex. 2.** $x + y = 5$, both 0, 1, 2, 3, 4, 5. **Ex. 3.** $R = \{(x, y)|x \geqslant 0, y \geqslant 0, 2x + y \leqslant 10\}$; 0–5, 0–10. **Ex. 4.** 0, $\frac{1}{4}$, 4, $-\frac{1}{8}$, $\frac{x}{x+3}$. **Ex. 5.** 11, 2, 6, $2\frac{1}{4}$, $4x^2 + 2$; $f(x) \geqslant 2$. **Ex. 6.** 0, $\frac{1}{2}$, $\sqrt{\{x(1-x)\}}$; $\sqrt{\{x(1-x)\}}$; $1 \leqslant x \leqslant 2$. **Ex. 7.** (i) $y = \frac{3-x}{2}$, (ii) $y = \frac{1}{x}$, (iii) $y = \frac{1-x}{1+x}$. **Ex. 8.** $y = x + 1$, $x + y = 0$. **Ex. 9.** $x + y = 2$. **Ex. 11.** $\{(x, y)|x < 4$ and $y > 0$ and $y < x\}$. **Ex. 12.** $x = |y|$ or $x = \sqrt{(y^2)}$. **Ex. 14.** (i) Yes, no, (ii) $x \, f \, y$.

EXERCISE 4.1 (p. 63)

2. Yes, no. **3.** 1, 0, −3. **4.** Domain all x, range $y \geqslant 0$, (ii) all x, $y \leqslant 0$, (iii) all x, all y, (iv) $x > 0$ and $x < 0$, $y = 1$ and $y = -1$. **8.** (i) f, $y = x^2$; g, $y = x^3$, (ii) f, $y = x^2$; g, $y = \sqrt{x}$.

CHAPTER 5 (pp. 65–82)

Exercise 5.1 (p. 69)

1. 4. **2.** 3. **3.** 3. **4.** 4. **5.** 3. **6.** 2. **7.** 9. **8.** 3. **9.** $\frac{1}{8}$. **10.** $\frac{1}{9}$. **11.** 27. **12.** $\frac{1}{64}$. **13.** 81. **14.** 3. **15.** 12. **16.** $\frac{1}{8}$. **17.** $\frac{1}{2}$. **18.** 16. **19.** $\frac{1}{4}$. **20.** 32. **21.** $\frac{1}{4}$. **22.** 0. **23.** $3\sqrt{2}$. **24.** $\sqrt[3]{4}(\sqrt[3]{4}-1)$. **25.** $\sqrt{5}$. **26.** $\sqrt[3]{2}$. **27.** $\sqrt{2}$. **28.** $3\frac{2}{3}\sqrt{3}$. **29.** 2. **30.** 9. **31.** $\frac{1}{8}$. **32.** $\frac{1}{3}$. **33.** 0. **34.** $-\frac{1}{2}$. **35.** 2. **36.** -1 or 27. **37.** $\frac{1}{2}\left(1+\frac{1}{a}\right)$. **38.** 2.

Exercise 5.2 (p. 74)

1. (i) 0·6020, (ii) 0·6990, (iii) 0·7781, (iv) 0·9030, (v) 0·9542, (vi) 0·0791, (vii) 0·1761, (viii) 0·2040, (ix) 0·2552. **2.** 1. **3.** 2. **4.** 0. **5.** 0. **6.** -1. **7.** 1. **8.** 3. **9.** $\frac{1}{2}$. **10.** 3.

Exercise 5.3 (p. 74)

1. (i) 6·29, (ii) -1·20. **2.** (i) 0·669, (ii) 0·808, (iii) 16·0. **3.** $y = 100x^{-2/3}$, 25.

Exercise 5.4 (p. 75)

1. 3. **2.** 2. **3.** $1\frac{1}{2}$. **4.** -2. **5.** -5. **6.** -5. **7.** 2. **8.** -3. **9.** -3.

Exercise 5.5 (p. 78)

1. (i) 7·2, (ii) 25·2, (iii) 1·93, (iv) 0·635, (v) 56·3(5), (vi) 1·11, (vii) 4·33, (viii) 3·98, (ix) 40·1. **2.** (i) 3·35 in., (ii) 11·30 in.; 1·68 in., 5·65 in. **3.** (i) 0·164 cm, (ii) 0·164 cm. **4.** (i) 61·5 cm, (ii) 34·0 cm, (iii) 63·4 cm. **5.** 2·55, 3·53; 28·4 cm.

Exercise 5.6 (p. 81)

1. (i) 0·5, (ii) 0·72. **2.** (i) 2·87, (ii) 0·64. **3.** (i) 0·37, (ii) 0·69.

Exercise 5.7 (p. 81)

1. 32. **2.** $\frac{1}{6}$. **3.** 6. **4.** $\frac{1}{32}x^{10}$. **5.** $\frac{1}{3}x$. **6.** $\frac{27}{x}$. **7.** $2\log 2 - 1$. **8.** 0. **9.** $1\frac{1}{2}$. **10.** $\frac{1}{2}$. **11.** 2. **12.** 6. **13.** 0 or 2. **14.** 1 or 2. **15.** 0. **16.** $1\frac{2}{3}$. **17.** 50. **18.** $\frac{1}{40}$. **19.** $\frac{1}{99}$. **20.** (2, 2). **21.** (1, 0). **22.** 4. **23.** 4. **24.** -2. **25.** $2\frac{1}{2}$. **26.** 0·966. **27.** 1·19. **28.** 0·0890. **29.** 0·298. **30.** 2·32. **31.** 1·77. **32.** -0·661. **33.** 0·906. **34.** 47·0. **35.** 0·482. **36.** 2·04. **37.** (i) 0·5, (ii) -2·0 or 0·7. **38.** (i) 1·4, (ii) 0·4, 3·5; 10^{-10}, 10. **39.** 1·98. **40.** 1996.

CHAPTER 6 (pp. 83–96)

Ex. 2. (i) 5, (ii) 5, (iii) 13, (iv) 25. **Ex. 4.** $7x + y = 8$. **Ex. 5.** $(x-2)^2 + (y-1)^2 = 9$. **Ex. 7.** (i) $\frac{1}{3}$, (ii) $-\frac{2}{3}$, (iii) $\frac{3}{5}$. **Ex. 11.** (i) 0, (ii) ∞, **Ex. 12.** (i) 2, (ii) $-\frac{2}{3}$, (iii) $\frac{1}{2}$, (iv) 0, (v) $\frac{3}{2}$, (vi) ∞. **Ex. 13.** (i) $2x - y = 4$,

(ii) $2x + 3y = 5$, (iii) $y = 2$, (iv) $x = 3$. **Ex. 15.** (i) $3x - y = 7$, (ii) $4x + 3y = -5$, (iii) $2x - 3y = -8$, (iv) $2x + 3y = 0$.
Ex. 16. (i) $x - 3y = -7$, (ii) $4x + 3y = 5$, (iii) $x + 3y = 3$, (iv) $x - y = 0$.
Ex. 18.(i) 4/5, (ii) 4/5, (iii) $\sqrt{2}$, (iv) 1. **Ex. 19.** (i) (3, −2), (ii) (0 −2).
Ex. 20. −2.

Exercise 6.1 (p. 95)

3. $\sqrt{3}x + y = 5\sqrt{3}$. **4.** (7, 0), (3, 1). **5.** (i) $x + 2y = 3$, (ii) (9/5, 3/5). **6.** $4\frac{1}{12}$, $2x = 3y$. **7.** $x - 3y + 1 = 0$, $5x - 2y + 5 = 0$, $4x + y = 22$. **8.** (2, 1), (4, −2), (0, −1/3), $(4, 2\frac{1}{3})$. **9.** $3x + y = 5$, $(\frac{1}{2}, 3\frac{1}{2})$. **11.** $8x + 6y = 9$, **12.** $3x + 7y = 34$ 4/23. **13.** $x + 7y = 28$, $8x - y = 15$, $7x - 8y = -13$. **14.** 43, −87.

CHAPTER 7 (pp. 97–)108

Ex. 1. $6\frac{1}{2}$, $5\frac{1}{2}$, $28\frac{1}{2}$. **Ex. 2.** $2{\cdot}2x^2 - 2{\cdot}6x - 5$. **Ex. 3.** $2\frac{1}{2}x - 5$.
Ex. 4. (i) $1{\cdot}4(x + 2)$, (ii) $3(x - \frac{1}{2})$. **Ex. 6.** (i) $2x + 1$, (ii) $-\dfrac{8x + 1}{3}$.
Ex. 7. (i) $2x + 5y = 31$, (ii) $7x - 2y + 17 = 0$. **Ex. 8.** (i) Plane bisecting angle between xOy, zOy planes, (ii) Plane bisecting angle between xOy, xOz planes, (iii) Plane parallel to xOy plane, through (0, 0, 1). **Ex. 9.** Sphere, centre O, radius 1. **Ex. 10.** (i) $x = \lambda$, $y = 2\lambda$, $z = 3\lambda$. (ii) $x = 1 + \lambda$, $y = 2 - \lambda$, $z = 1 + \lambda$, (iii) $x = 2\lambda - 1$, $y = 2 - 4\lambda$, $z = 0$, (iv) $x = 3 - 4\lambda$, $y = 4\lambda - 1$, $z = 3\lambda + 2$. **Ex. 11.** (i) A line, (ii) a plane containing (i). **Ex. 14.** 4·4, $(4\frac{9}{23})$. **Ex. 15.** 11·8, $(11\frac{16}{19})$. **Ex. 16.** 21. **Ex. 17.** 15·6.

Exercise 7.1 (p. 108)

1. 2, 1, 1, 1, −2, −1, −1, −1. **2.** $2x = y$ and $x + 2y = 5$. **3.** (3, 3), 5. **4.** A closed hemisphere. **6.** 7 sq. units. **7.** 1·5 kg. **8.** 9 kg. **9.** (i) \$20, 0, 0; (ii) \$120/7, \$160/7, 0; (iii) \$3000/186, \$4000/186, \$5000/186 per child. **10.** 5800 fr. **11.** $(\lambda a, \lambda b, 6 - 6\lambda)$, $z = 6 - 6\lambda$. **12** (i) $4x + 3y - z = 12$, (ii) $4x + 3y + 2z = 12$; $4x + 3y = 12$, $z = 0$. **13.** 0, 0, −12, 24. **14.** $\left\{z^2 + \dfrac{(x - 2y)^2}{5}\right\}^{1/2}$. Inside of an endless cylinder, axis $x - 2y = 0$, $z = 0$, radius 1.

CHAPTER 8 (pp. 109–27)

Ex. 1. (i) $x = -1$, (−1, 9), (ii) $x = \frac{1}{2}$, $(\frac{1}{2}, 0)$, (iii) $x = \frac{3}{4}$, $(\frac{3}{4}, \frac{7}{8})$.
Ex. 2. (i) $x < -2$ and $x > 3/2$, (ii) $-2/3 < x < 5/2$.

Exercise 8.1 (p. 111)

1. (i) $x < 1$, $x > 2$, (ii) $1 < x < 2$. **2.** (i) $x < -2$, $x > 0$, (ii) $-2 < x < 0$. **3.** (i) $-2 < x < 1$, (ii) $x < -2$, $x > 1$. **4.** (i) $x < -\frac{3}{2}$, $x > 1$, (ii) $-\frac{3}{2} < x < 1$. **5.** (i) $-\frac{3}{2} < x < 2$, (ii) $x < -\frac{3}{2}$, $x > 2$. **6.** (i) All except $x = 3/2$, (ii) No value. **7.** +ve, 2, −1. **8.** −ve, −1, 2. **9.** +ve, 4, $-\frac{1}{2}$. **10.** −ve, $-\frac{3}{4}$, $\frac{1}{2}$. **11.** −ve, −3, −5/2. **42.** +ve, $\frac{1}{4}$, $\frac{1}{4}$. **13.** Only −ve for $-2 < x < -\frac{1}{2}$. **11.** +ve, but 0 when $x = -1$. **15.** +ve. Least value 7/8 when $x = -\frac{3}{4}$.

16. −ve. Greatest value −5 when $x = -\frac{1}{2}$. **17.** +ve for $-\frac{1}{2} < x < 5$. **18.** −ve, but 0 when $x = 5/3$. **19.** $x < 2$ or $x > 5$. **20.** $-1 < x < 3$. **21.** $x < -3$ or $x > 2$. **22.** $-2 < x < 1$. **23.** $x < -3$ or $x > -2$. **24.** $-3/2 < x < 3/2$.

Exercise 8.2 (p. 115)

2. $-3{\cdot}2 < x < 2{\cdot}2$; (i) $-3 < x < 2$, (ii) $-2 < x < 1$, (iii) $-3{\cdot}05 < x < 2{\cdot}05$. **3.** $-3/2$, $1/4$. **4.** 3, 1. **5.** −1, 1. **7.** $x < -3$ or $x > 2/3$; $10\frac{1}{12}$. **9.** $< -4 - 2\sqrt{3}$ and $> -4 + 2\sqrt{3}$. **10.** $0 < k < 6$. **11.** $4(x - 1)(3 - x)$. **12.** $3(x + 2)(x - 1)$. **13.** $a < -\frac{4}{3}$ or $a > 0$. **15.** $1 < x < 9/5$; −1, −4.

Ex. 3. (i) 8, (ii) 14, (iii) $-28\sqrt{2}$, (iv) 136. **Ex. 4.** (i) $2x^2 + 13x + 19 = 0$. (ii) $2y^2 - 21y + 2 = 0$.

Exercise 8.3 (p. 121)

1. $3\frac{3}{4}$, $2\frac{3}{4}$. **2.** $C = b^2 - 1$. **3.** (i) $-\alpha, -\beta$; (ii) $\frac{1}{\alpha}, \frac{1}{\beta}$: $a^2x^2 + ax(b - c) - bc = 0$.

5. $kx^2 - 3x(k + 1) + (k + 1)^2 = 0$. $\frac{1}{k^2}(9 - 2k)(1 + k)^2$.

7. $x^2 - 21x + 4 = 0$. **8.** $a \leqslant 10\frac{1}{8}$. **9.** $9x^2 - 48x + 4 = 0$. **10.** −4, 9. **11.** $4x^2 + 33x + 4 = 0$. **12.** $x^2 + x\{2(c - 1) - b^2\} + (c - 1)^2 + b^2 = 0$. **13.** $3b$. **14.** $x^2 - x(p + 2)^2 + 4p(p^2 + 4) = 0$. **16.** −2, $-4 \leqslant p \leqslant 4$.

Exercise 8.4 (p. 124)

2. $1 \pm j$. **3.** $-1 \pm 2j$. **4.** $-2 \pm 2j$. **5.** $-1 \pm j\sqrt{3}$. **6.** $3 \pm 2j$. **7.** $-2 \pm j\sqrt{5}$. **8.** $\frac{1}{2}(1 \pm j)$. **9.** $\frac{1}{2}(3 \pm j\sqrt{3})$. **10.** $\frac{1}{3}(2 \pm j\sqrt{3})$. **11.** $5 + 4j$. **12.** $2(1 + j)$. **13.** $4(1 + j)$. **14.** $-(1 + j)$. **15.** $4 - 3j$. **16.** $-1 + 3j$. **17.** $3 + 4j$. **18.** $-3 - 4j$. **19.** $\frac{1}{5}(1 + 3j)$. **20.** $\frac{1}{2}(1 - 3j)$. **21.** $\frac{1}{5}(-1 + 8j)$. **22.** $\frac{1}{13}(2 + 3j)$. **23.** $\frac{p - jq}{p^2 + q^2}$. **24.** $\frac{p(-q + jp)}{p^2 + q^2}$. **25.** $\frac{(ap + bq) + j(aq - bp)}{a^2 + b^2}$.

26. (i) 2, $11j$; (ii) 2, j. **27.** (i) $3 - 2j$, $\sqrt{13}$, (ii) $-8 - j$, $\sqrt{65}$, (iii) $\frac{-3 - j}{5}$, $\frac{1}{5}\sqrt{10}$.

Exercise 8.5 (p. 127)

1. [2, ±3]. **2.** [−3, 0] or [7, 0]. **3.** [0, ±2]. **4.** [−5/3, ±1]. **5.** $[0, \pm\frac{4}{3}\sqrt{3}]$. **6.** $[\frac{1}{3}(-2 \pm \sqrt{46}), 0]$. **7.** [2, −1], [−1, 5], [4, −3], [−5/2, −1/2]. **8.** [−1, 5], [−12, −5], [−5, −1], [0, 1]. **9.** [6, 0], [13, 0], [0, −4], $[\frac{5}{13}, -\frac{12}{13}]$. **10.** [−1, 5], [−4, −4], [1, 3], [2, −2]. **11.** [1, 1], [−2, −1], [3, 1], [−2, −1]. **12.** [0, 0], [21, 20], [−4, 10], [−1, 0]. **13.** $[-1, -1]$, $[\frac{1}{2}, -\frac{1}{2}]$. **14.** [−2, 1], [2/5, 1/5]. **15.** [2, −1], [−2/5, −1/5]. **16.** [0, −1], [0, −1]. **17.** [1, 0], [−1, 0]. **18.** [−3, 4], [3/25, 4/25].

CHAPTER 9 (pp. 128–40)

Exercise 9.1 (p. 128)

14. $3r - 2$. **15.** 4^{r-1}. **16.** $(-1)^{r+1}/r$. **17.** $7 - 2r$. **18.** $-(-2)^r$.
19. $(2r)^2$. **20.** $2^r - 1$. **21.** $1 - 2^{1-r}$. **22.** $r - 1/r$.

Ex. 1. $4n - 3$; $n(2n - 1)$. **Ex. 4.** $8\frac{1}{3}$, $5\frac{2}{3}$. **Ex. 5.** (i) -1, 15th, (ii) 29. **Ex. 6.** -2; $2\sqrt{2}$. **Ex. 7.** 4, 2, 1, $\frac{1}{2}$.
Ex. 8. (iii) $\frac{2}{45}\{1 - (1{\cdot}5)^{2n}\}$, (iv) $\frac{2}{45}\{1 + (1{\cdot}5)^{2n+1}\}$.

Exercise 9.2 (p. 134)

1. A.S.; $3n + 1$; $\frac{1}{2}n(3n + 5)$. **2.** G.S.; $3/2^{n-3}$; $24(1 - 1/2^n)$. **3.** No; n^2.
4. G.S.; 3^{n-1}; $\frac{1}{2}(3^n - 1)$. **5.** A.S.; $11 - 4n$; $n(9 - 2n)$.
6. G.S.; $(-2)^n$; $\frac{2}{3}\{(-2)^n - 1\}$. **7.** G.S.; $\frac{1}{3}(2^{n-5})$; $\frac{1}{48}(2^n - 1)$.
8. A.S.; $n/48$; $\frac{1}{96}n(n + 1)$. **9.** No; $1/(3n - 1)$. **10.** 3367. **11.** 59.
12. (i) 28, (ii) 29, 260. **13.** (i) $r + 1$, (ii) $(r + 1)(2r + 1)$
14. (i) 46. 39, 32, (ii) $15a - 13b$, $13a - 10b$, $11a - 7b$.
15. (i) $6\frac{1}{3}$, (ii) $\frac{1}{3}(13x + 36)$. **16.** (i) 320, 160, 80, (ii) $\frac{5}{3}$, -5, 15.
18. (i) 8, (ii) 7, (iii) 9, (iv) 6, (v) 5, (vi) 34. **19.** (i) \$871.25, (ii) \$1087.
20. \$2457, \$1282. **21.** £1478. **22.** \$3100.

Ex. 9. 12; 16. **Ex. 10.** (i) 1, (ii) $\frac{1}{2}$, (iii) 2. **Ex. 12.** 0.

Exercise 9.3 (p. 140)

1. (i) $2 - 2^{1-n}$; 2, (ii) $\frac{9}{4} - \frac{1}{4}(-3)^{2-n}$; $\frac{9}{4}$, (iii) $3 - 3(\frac{2}{3})^n$; 3. **2.** (i) $\frac{2}{3}$, (ii) $\frac{3}{4}$, (iii) 0, (iv) 0, (v) 1, (vi) $\frac{1}{3}$. **3.** $|x - 2| < 0{\cdot}19$; 3. **4.** $\frac{1}{2}$.
5. $\dfrac{1 - 2\epsilon}{1 + 2\epsilon} < x < \dfrac{1 + 2\epsilon}{1 - 2\epsilon}$.

CHAPTER 10 (pp. 141–54)

Ex. 1. (i) $\dfrac{n}{n + 1}$, (ii) n^3, (iii) $\frac{1}{4}n^2(n + 1)^2$. **Ex. 2.** (i) 9, (ii) 64.
Ex. 3. $n! > 2^{2n}$ when $n > 8$.

Exercise 10.1 (p. 146)

7. 9. **16.** (i) 8, (ii) 2. **17.** (i) $n + 1$, (ii) $2n - 1$. **18.** 7.
20. $2(3^{n-1} - 1)$. **23.** $n^2 - n + 2$. **24.** $AB/(n + 1)$.

Ex. 6. (i) $1 + \frac{1}{2} + \frac{1}{3} + \frac{1}{4}$, (ii) $3 \cdot 4 + 4 \cdot 5 + 5 \cdot 6$, (iii) $1 - 2 + 4 - 8 + 16$.

Ex. 7. (i) $\sum_{r=1}^{4} r^2$, (ii) $\sum_{r=2}^{6} (-1)^r \cdot \dfrac{1}{r}$ or $\sum_{r=1}^{5} (-1)^{r+1} \dfrac{1}{r + 1}$, (iii) $\sum_{r=1}^{2n} (-1)^{r+1} r^3$.

Ex. 8. $\frac{1}{4}n(n + 1)(n + 2)(n + 3)$. **Ex. 9.** $\dfrac{1}{18} - \dfrac{1}{3(n + 1)(n + 2)(n + 3)}$.

Ex. 10. $\frac{1}{4}n^2(n + 1)^2$. **Ex. 11.** $\dfrac{11}{18} - \dfrac{1}{3}\left(\dfrac{1}{n + 1} + \dfrac{1}{n + 2} + \dfrac{1}{n + 3}\right)$.

EXERCISE 10.2 (p. 153)

1. $\frac{3}{2}n(n+1)$. **2.** $\frac{1}{6}n(n+1)(2n+7)$. **3.** $2^{n+1}-(n+2)$. **4.** $\frac{3}{4}-\frac{2n+3}{2(n+1)(n+2)}$. **5.** $2n-1$. **6.** $9n^2-5n+1$. **7.** $2 \cdot 3^{n-1}+1$, $(n>1)$. **8.** $\log\frac{n+1}{n}$. **9.** (i) $\frac{1}{6}n(n+1)(2n+7)$, (ii) $\frac{1}{6}n(n+1)(3n^2+5n+1)$, (iii) $\frac{1}{3}n(4n^2+12n-1)$. **11.** $\frac{1}{3}n(4n^2+6n-1)$; $\frac{4}{3}n(n+1)(n+2)$. **12.** $n/(2n+1)$. **13.** $\frac{1}{3}n(4n^2+18n+23)$.

14. $\frac{1}{6}n(n+1)(4n+11)$. **15.** $n(n+1)(n-2)$. **16.** $\frac{n}{3n+1}$.

17. $\frac{n(4n+5)}{3(2n+1)(2n+3)}$. **18.** $\log \frac{1}{2}(n+1)(n+2)$. **19.** (i) $\frac{1}{6}$; (ii) 1; (iii) $\frac{5}{4}$. **20.** $\sum_{r=1}^{n} r(r+1)(r+2)\ldots(r+k-1)$ is a polynomial in n of degree $(k+1)$ which is always divisible by $k!$.

CHAPTER 11 (pp. 155–69)

Ex. 1. (i) 5, 4·1, $4+h$, (ii) 3, 3·9, $4-h$.

EXERCISE 11.1 (p. 158)

1. (i) 3, (ii) -1, (iii) 1·5, (iv) -5. **2.** (i) 3, (ii) 2·5, (iii) 2·2, (iv) 2·1, (v) 2·01, (vi) 2·001. **3.** (i) $32\frac{1}{3}$, (ii) 39, (iii) $42\frac{1}{3}$, (iv) 35·3 ft/sec. **4.** (i) $-1/3$, (ii) $-1/2$, (iii) $-2/3$, (iv) $-10/11$, (v) $-100/101$.

Ex. 2. (i) $4x$, (ii) 3, (iii) $2x+5$.

EXERCISE 11.2 (p. 164)

1. $6x$. **2.** $3x^2$. **3.** $8/(3-2x)^2$. **4.** $4x^3$. **5.** $-2/x^3$. **6.** $-4/(x+1)^3$. **8.** $x=y$. **9.** (i) $5x-y=3$, (ii) $x+8y=8$. **10.** (2, 3), $(\frac{1}{3}, -5\frac{1}{3})$. **11.** (i) $(\frac{1}{2}, -2)$, $(-\frac{1}{2}, 2)$, (ii) $(\frac{1}{2}, 0)$. **12.** (i) 17, (ii) 42 m/sec.

Ex. 3. (i) $8x^7$, (ii) $\frac{3}{2}\sqrt{x}$, (iii) $-1/(2\sqrt{x^3})$, (iv) $1/(3\sqrt[3]{x^2})$, (v) $-\frac{5}{x^6}$.

Ex. 5. (i) $4(x+2)^3$, (ii) $\frac{3}{x^4}(x^2+1)^2(x^2-1)$, (iii) $\frac{3x^2}{2\sqrt{(x^3+1)}}$, (iv) $3\sqrt{(2x-3)}$.

EXERCISE 11.3 (p. 169)

1. (i) $9x^2$, (ii) $4-10x$, (iii) $10x-8x^3$, (iv) $1/x^2$, (v) $-2/x^2+6/x^3$, (vi) $(10-4x)/x^3$, (vii) $2/\sqrt{x}$, (viii) $3/(x^2\sqrt{x})$, (ix) $(\sqrt{x}-4)/(2x^2)$. **2.** (i) $8(2x+3)^3$, (ii) $3/2\sqrt{(3x-4)}$, (iii) $3/(5-3x)^2$, (iv) $18x(3x^2+2)^2$, (v) $x/\sqrt{(x^2-5)}$, (vi) $4x/(3-2x^2)^2$, (vii) $9(x-1)/2(3x-2)^{3/2}$, (viii) $x(4-x^2)/(x^2+2)^{5/2}$, (ix) $4x(1-2x^2)/(3-2x^2)^3$. **3.** (i) $-4/3$, (ii) $9/4$, (iii) $-3/2$, (iv) $-\frac{1}{2}$. **4.** (i) $x+y=3$, (ii) $x+4y=9a$, (iii) $3x-y=9$, (iv) $3x+2y+8=0$. **5.** (i) $x-y+1=0$, (ii) $4x+3y=\pm 25$. **6.** 7/12 ft/sec. **7.** $x+a^2y=2a$.

CHAPTER 12 (pp. 170–84)

EXERCISE 12.1 (p. 171)

1. 5π. **2.** 6π cm/sec, 9π cm²/sec. **3.** -25. **4.** Decreasing at 0·08 per sec. **5.** (i) $\frac{1}{20\pi}$ cm/sec, (ii) 5 cm³/sec. **6.** 5/18 cm/sec; 18 cm². **7.** $1/t$. **8.** $3t/2$. **9.** $-1/t^2$. **10.** $\frac{3-2t}{2t}$.
11. $4t\sqrt{(1+t)}$. **12.** $-\frac{1}{t^3}\sqrt{(1-t^2)^5}$.
Ex. **1.** (i) $(-1, 3)$ relative maximum; $(1, -1)$ relative minimum, (ii) $(-2, -\frac{1}{4})$, min.

EXERCISE 12.2 (p. 175)

1. (i) $x > 2$, (ii) $x < 2$; 1, when $x = 2$, min. **2.** (i) $x < 1$, $x > 2$, (ii) $1 < x < 2$; 1, when $x = 1$, rel. max.; 0, when $x = 2$, rel. min. **3.** (i) 14, when $x = 2$, max., (ii) -13, when $x = 4$, min., (iii) -4, when $x = 1$, min., (iv) $\sqrt{2}$, when $x = 0$, max., (v) $\sqrt{\frac{1}{2}}$, when $x = 0$, min., (vi) -2 when $x = -1$, rel. max., 2, when $x = 1$, rel. min. **4.** (i) 5, 5 cm, (ii) 0,10 cm, (iii) 8,2 cm. **5.** A square. **6.** $12\frac{1}{2}$ cm². **7.** $12\frac{1}{2}$ cm². **8.** 10π cm. **9.** $\frac{1}{2}$. **10.** $5\sqrt{5}$.

EXERCISE 12.3 (p. 177)

1. π cm². **2.** 80π cm³. **3.** 3%. **4.** 8·4%. **5.** $+\frac{1}{4}$%. **6.** 0·11 dynes. **7.** $-v^2/u^2$. **8.** $1{\cdot}1\pi$.
Ex. **2.** (i) $y = x(x^3 - 3x + 3)$, (ii) $\frac{3}{2}x^2 + \frac{2}{x} - \frac{1}{2x^2} - \frac{7}{8}$.

EXERCISE 12.4 (p. 180)

2. (i) $4x$, (ii) $\frac{4}{3}x^3$, (iii) $\frac{4}{3}x^2 + 2x$, (iv) $\frac{1}{5}x^5$, (v) $2/x$, (vi) $x^{-1} - x^{-2}$, (vii) $\frac{1}{3}x^3 - x^{-1}$, (viii) $2x^{3/2}$, (ix) $\frac{1}{3}x^3 - x^2 + 3x$, (x) $\frac{2}{3}(x+3)\sqrt{x}$, (xi) $\frac{2}{3}(2x-3)\sqrt{x}$, (xii) $x^{-2} - 3x^{-1}$. **3.** (i) $y = \frac{1}{9}x^3 + \frac{1}{4}x^2 + \frac{1}{2} + 3x^{-1}$, (ii) $y = 2x^{5/2} - 2x^{3/2} + 4x - 36$. **4.** $y = x^3 - 4x^2 + 4x + 5$. **5.** -4. **6.** 3150 ft. **7.** 213 ft. **8.** $y = \frac{1}{10}(x^3 + 3x^2 - 24x + 10)$; $(-4, 9)$, $(2, -9/5)$. **9.** 6·75 cm; 4 cm/sec.

EXERCISE 12.5 (p. 183)

1. 112·5 m. **2.** 10·5 cm/sec; $13\frac{1}{6}$ cm. **3.** ± 5 m/sec; 2·5. **4.** $16\sqrt{2}$ft/sec; 32 ft. **5.** (i) -10 ft/sec²; (ii) $-125/27$ ft., (iii) 3 ft/sec. **6.** 1 sec,7 sec; 2·1 sec. **7.** (i) 75 ft/sec., (ii) 250 ft; 20. **8.** (ii) 18 sec; 27 sec. **9.** (i) $-c^2/2a^2$ km/hr², (ii) $2(a^{3/2} - b^{3/2})/3c$ hr. **10.** $2kx^{3/4}$ cm/sec, $2x^{1/4}/k$ sec.

CHAPTER 13 (pp. 185–200)

EXERCISE 13.1 (p. 188)

2. $(1 - x^2)/(x^2 + 1)^2$. **3.** (i) $2x(x-1)(2x-1)$, (ii) $(3x+2)/2\sqrt{(x+1)}$, (iii) $2(x+2)(x-1)(2x+1)$, (iv) $5(2x-1)(6x-1)(3x+1)^2$, (v) $(4x^2 + x + 4)/\sqrt{(x^2+2)}$, (vi) $2n(1-x)(2x-x^2)^{n-1}$. **4.** (i) $12/(2-3x)^2$,

(ii) $2(x + 1)\quad(x - 2)/(2x - 1)^2$, (iii) $-(2 + x)/2x^2\sqrt{(1 + x)}$, (iv) $-(ad + bc)/(c + dx)^2$, (v) $-2/x^2\sqrt{(x^2 + 2)}$, (vi) $\frac{1}{2}x^{-1/2}(x + 1)^{-3/2}$. **6.** Min., $-\frac{1}{6}$ when $x = -2$; max., $-\frac{1}{3}$ when $x = 1$. **7.** $11x + 2y - 24 = 0$, $7x - y - 12 = 0$. **9.** $5x - 4y - 12 = 0$, $4x + 5y - 26 = 0$. **12.** $x^3 = Ax + 2y$. **13.** $(x + y)^2 = x^4 + Ax^2$. **14.** (i) Max., $\frac{4}{27}$, when $x = \frac{4}{3}$; min., 0, when $x = 2$, (ii) Max., $3 - 2\sqrt{2}$ when $x = 1 - \sqrt{2}$; min., $3 + 2\sqrt{2}$, when $x = 1 + \sqrt{2}$, (iii) Max., $\frac{1}{8}\sqrt{2}$, when $x = \frac{1}{2}$; min., $\frac{2}{3}\sqrt{3}$, when $x = -2$, (iv) None.

Exercise 13.2 (p. 195)

1. (i) Max., 27, $x = -2$; min., 0, $x = 1$; infl., $(-\frac{1}{2}, \frac{27}{2})$, (ii) Max., 327, $x = 3$; min., -16, $x = -\frac{1}{2}$; infl., $(\frac{5}{4}, \frac{311}{2})$, (iii) Min., -54, $x = -3$; max., 10, $x = -1$; min., -179, $x = 2$; infl., $x = \frac{1}{3}(-2 \pm \sqrt{21})$, (iv) Min., 0, $x = -2$; max., $\frac{8}{8}$, $x = -\frac{3}{2}$; min., $-\frac{125}{8}$, $x = \frac{1}{2}$; infl., $x = \frac{1}{6}(-6 \pm \sqrt{19})$. **2.** (i) $y = 2x^3 - 3x^2 - x + 1$, (ii) $y = 7 - 17x + 6x^2 + 3x^3 - x^4$, (iii) $y = x/(x^2 + 1)$; $(-\sqrt{3}, -\frac{1}{4}\sqrt{3})$, $(0, 0)$, $(\sqrt{3}, \frac{1}{4}\sqrt{3})$. **3.** Max., (1, 1); min., $(-4, -1/4)$. **4.** $(1 + 2t)^2/t^2$; $2(1 + 2t)^3/3t^3$. **5.** $-2t/(t + 1)$. **6.** $(3t^2 - 1)/2t$; $-(3t^2 + 1)/4t^3$.

Exercise 13.3 (p. 199)

1. $v^2 = 2as$, $v = at$; 24 ft. **2.** (i) 12 ft/sec, (ii) $13\frac{2}{3}$ ft. **3.** 12 ft/sec², $t = 2$; -12 ft/sec², $t = -2$; ± 12 ft/sec². **4.** -1. **5.** $x = y^2 + 3y - 1$. **6.** $-(x + y)/(x + 3y)$. **7.** $(t^2 + 1)/(t^2 - 1)$, $-4t^3/(t^2 - 1)^3$. **8.** $-5t^{3/2}/6(1 + t^2)$, $-5t^{1/2}(3 - t^2)/36(1 + t^2)^3$; $0, \pm 6\sqrt{3}$. **9.** $y = x^3 + \frac{3}{2}x^2 - 18x + 3$. **10.** $10\sqrt{2}/3$.

CHAPTER 14 (pp. 201)

Ex. 3. (i) $x^3 + x^{-1} + C$, (ii) $\frac{1}{8}(2x + 3)^4 + C$, (iii) $-\frac{3}{2}(2x - 1)^{-1} + C$. **Ex. 4.** (i) 4, (ii) 11/5, (iii) 15/64, (iv) 45/16. **Ex. 5.** $\frac{3}{4}$. **Ex. 6.** $-32/3$. **Ex. 7.** $14\frac{1}{2}$, 8. **Ex. 8.** 14/3.

Exercise 14.1 (constants of integration omitted) (p. 211)

1. (i) $\frac{3}{4}x^4 - x^2 + x$, (ii) $\frac{1}{12}(2x - 5)^6$, (iii) $-\frac{1}{9}(3x + 1)^{-3}$, (iv) $\frac{2}{3}(x + 6)^{3/2}$, (v) $\frac{3}{4}(2x + 3)^{2/3}$, (vi) $\frac{2}{3}x^{3/2} - 2x^{1/2}$. **2.** (i) 4, (ii) 7, (iii) 195/32, (iv) $\frac{2}{3}(3\sqrt{3}-1)$, (v) 3724/3, (vi) $1/3\sqrt{3}$. **3.** (i) $\sqrt{(x - 1)}$ does not exist when $0 \leqslant x < 1$, (ii) there is an asymptote $x = 0$, (iii) as in (ii), (iv) $\sqrt{(1 + x^2)}$ cannot be integrated at this stage. **4.** (i) $2\int_0^1 (x^2 + 1)dx$, (ii) 0, (iii) $2\int_0^1 x^2 dx$. **5.** (i) 7·5, (ii) 22·5. **6.** (i) $-4/3$, (ii) 125/216, (iii) $-2/3$, (iv) 343/54. **7.** $-1/6$. **8.** 4·5. **9.** 343/6. **10.** 25/3· **11.** 12. **12.** $32\sqrt{2}/3$. **13.** 125/6. **14.** $1 - X^{-1}$; $1 \,.\, x_1^{-1} - 1$; ∞· **15.** 4. **16.** 27/4.

Ex. 10. $\frac{1}{3}\pi(q^3 - p^3)$; the portion of a right circular cone with its axis along OX and vertical angle a right angle, lying between the planes $x = p$, $x = q$.

Exercise 14.2 (p. 215)

1. $20\pi/3$. **2.** 93π. **3.** $4\pi a^3$. **4.** $152\pi/3$. **5.** $\frac{1}{3}\pi h^2(3a - h)$.
6. (i) 16π, (ii) 24π. **7.** $512\pi/15$. **8.** (i) 59/12, (ii) $162\pi/35$.
9. 16π. **10.** $52000\pi/3$.

Ex. 12. $(\frac{3}{5}b, \frac{3}{4}\sqrt{ab})$; $(\frac{2}{3}b, 0)$. **Ex. 13.** (i) 0·6605, (ii) 0·6641, (iii) 0·6667.
Ex. 14. 0·1678.

Exercise 14.3 (p. 223)

1. 38/15. **2.** 2 ft/sec. **3.** 500/211 g/cm^2. **4.** (i) $65\pi/3$, (ii) $65\pi/3$.
5. (i) 350 ft/sec, (ii) 1400/3 ft/sec. Part (i) gives the same result as (total distance)/time. **6.** $(\frac{4}{7}b, \frac{5}{8}a^{2/3}b^{1/3})$; $(\frac{5}{8}b, 0)$. **7.** $\frac{3}{8}a$ from the centre.
8. (i) one-quarter of the way up the central axis, (ii) one-third of the way up.
9. 25/8. **10.** $h(a^2 + 2ab + 3b^2)/4(a^2 + ab + b^2)$. **11.** 28·0 cm^2; 3·11.
12. (38/9, 0). **13.** 57·18, 57·30. **14.** 0·6938, 0·6931.

INDEX

MODERN MATHEMATICS IN SECONDARY SCHOOLS

D. T. E. MARJORAM, B.Sc., Dip. Ed., A.F.I.M.A.
Deputy Headmaster, Doncaster Technical Grammar School for Boys

In providing a simple introduction to most of the topics which have found, or are finding, their way into revised syllabuses in school mathematics—and in discussing ways in which these subjects might be treated in the classroom—the author takes a completely new look at familiar concepts. Topics covered—most of which are new to Secondary School mathematics syllabuses—include sets, applications of set theory and Boolean algebra, groups, matrices, vectors, linear programming, statistics and patterns in arithmetic. Each topic is developed out of preceding ones and numerous worked examples are given. Each chapter concludes with a list of the latest relevant publications for further reading, and the appendix contains a variety of exercises on the whole book together with answers, hints and, in many cases, full solutions.

The book will be of great value to teachers in that it will undoubtedly help those faced with the task of preparing pupils for any of the new syllabuses in 'O' level mathematics. Teachers in training will also find it useful, and it may be used by sixth form pupils as supplementary reading.

EXERCISES IN MODERN MATHEMATICS

D. T. E. MARJORAM, B.Sc., Dip. Ed., A.F.I.M.A.
Deputy Headmaster, Doncaster Technical Grammar School for Boys

All the demands of the various new 'O' level syllabuses in mathematics are covered by this volume, used in conjunction with the normal school textbooks, It is written for secondary school pupils and provides the teacher with a wide variety of exercises on Sets; Sets of Points; Linear Programming; Sets, Logic and Switching Circuits; Scales of Notation, Groups; Matrices; Vectors; Probability and Statistics; and simple Topology. Each topic is developed through a number of carefully graded exercises; the earlier ones in each chapter are suitable for 1st and 2nd year pupils and the final exercises are of approximately 'O' level standard.

A useful point is that there is sufficient explanation to enable the brighter pupil to proceed unaided. Although many students will use the book selectively, it possesses a unity of its own. Each chapter is developed from the previous one and older pupils might work straight through the book as an introduction to more advanced texts in modern mathematics. There are ample examples and an abundance of very carefully graded exercises; these latter are essentially practical, for most of them have been used in the author's school and where necessary have been revised in the light of that experience.

The book is a companion volume to the author's *Modern Mathematics in Secondary Schools*, which is written mainly for teachers and intending teachers.

Printed in Dunstable, United Kingdom

91341557R00141